621.38416
A777
v.2

621.38416 A777
The ARRL UHF/microwave
projects manual
50425

WITHDRAWN

HAWKEYE COMMUNITY COLLEGE
3 7944 1008 72

050425

The ARRL UHF/Microwave Projects Manual

Volume 2

D1296991

Published By The American Radio Relay League
225 Main Street
Newington, CT 06111

Copyright © 1997 by

The American Radio Relay League

Copyright secured under the Pan-American Convention

International Copyright secured

This work is publication No. 226 of the Radio Amateur's Library, published by the League. All rights reserved. No part of this work may be reproduced in any form except by written permission of the publisher. All rights of translation are reserved.

Printed in USA

Quedan reservados todos los derechos

ISBN: 0-87259-631-1

Acknowledgments

This book was made possible by the active cooperation and help of the authors of the various projects, who were willing to share their hard work with the rest of the ham community.

Paul Danzer, N1II, Joel Kleinman, N1BKE, and Zack Lau, W1VT, compiled the contents of this book. Shelly Bloom, WB1ENT, Dave Pingree, N1NAS, Sue Fagan, Paul Lappen, Joe Shea and Steffie Nelson designed, produced and proofread the book. Sue Fagan also designed the cover.

Foreword

The first volume of this book was published in 1994, and it continues to be in demand. Since then, microwave technology has invaded the lives of consumers as never before. It seems that every other person is carrying an 800-MHz radio, better known as a cellular phone. Wireless cable at 2 GHz continues to grow. Home satellite TV reception is becoming commonplace. Wireless LANs are running on frequencies throughout the microwave spectrum.

What does this mean to hams? More and more components are becoming available, many commercial parts and subassemblies are starting to appear in VHF/UHF and microwave equipment, and many hams are jumping into an area previously thought to be only for a few experimenters.

Although material from this book comes from many sources, there is one consistent theme—you can build it! It does not take a room full of special test equipment and machine tools to assemble a high performance antenna or transverter. The authors of the projects in this book will show you how.

After all is said and done, the real reason hams gravitate toward the UHF and microwave bands is all the interesting, fun things to do with them. With the help of the projects in this book, you'll be able to join in.

Our thanks to the many authors of this book who have generously provided their permission for us to compile their work into this volume. If you have any comments or suggestions—or if you have an idea for a project of your own design—we would appreciate hearing from you. There's a handy Feedback form at the back.

David Sumner, K1ZZ
Executive Vice President
Newington, Connecticut
October 1997

Contents

About the
American Radio Relay League

The seed for Amateur Radio was planted in the 1890s, when Guglielmo Marconi began his experiments in wireless telegraphy. Soon he was joined by dozens, then hundreds, of others who were enthusiastic about sending and receiving messages through the air—some with a commercial interest, but others solely out of a love for this new communications medium. The United States government began licensing Amateur Radio operators in 1912.

By 1914, there were thousands of Amateur Radio operators—hams—in the United States. Hiram Percy Maxim, a leading Hartford, Connecticut, inventor and industrialist saw the need for an organization to band together this fledgling group of radio experimenters. In May 1914 he founded the American Radio Relay League (ARRL) to meet that need.

Today ARRL, with more than 170,000 members, is the largest organization of radio amateurs in the United States. The League is a not-for-profit organization that:

- promotes interest in Amateur Radio communications and experimentation
- represents US radio amateurs in legislative matters, and
- maintains fraternalism and a high standard of conduct among Amateur Radio operators.

At League headquarters in the Hartford suburb of Newington, the staff helps serve the needs of members. ARRL is also International Secretariat for the International Amateur Radio Union, which is made up of similar societies in more than 150 countries around the world.

ARRL publishes the monthly journal *QST*, as well as newsletters and many publications covering all aspects of Amateur Radio. Its Headquarters station, W1AW, transmits Morse code practice sessions and bulletins of interest to radio amateurs. The League also coordinates an extensive field organization, which provides technical and other support for radio amateurs as well as communications for public service activities. ARRL also represents US amateurs with the Federal Communications Commission and other government agencies in the US and abroad.

Membership in ARRL means much more than receiving *QST* each month. In addition to the services already described, ARRL offers membership services on a personal level, such as the ARRL Volunteer Examiner Coordinator Program and a QSL bureau.

Full ARRL membership (available only to licensed radio amateurs in the US) gives you a voice in how the affairs of the organization are governed. League policy is set by a Board of Directors (one from each of 15 Divisions). Each year, half of the ARRL Board of Directors stands for election by the Full Members they represent. The day-to-day operation of ARRL HQ is managed by an Executive Vice President and a Chief Financial Officer.

No matter what aspect of Amateur Radio attracts you, ARRL membership is relevant and important. There would be no Amateur Radio as we know it today were it not for the ARRL. We would be happy to welcome you as a member! (An Amateur Radio license is not required for Associate Membership.) For more information about the ARRL and answers to any questions you may have about Amateur Radio, write or call:

ARRL
225 Main Street
Newington CT 06111-1494
860-594-0200
Prospective new amateurs call:
800-32-NEW HAM (800-326-3942)
E-mail: **newham@arrl.org**
World Wide Web: **http://www.arrl.org/**

ANTENNAS

Practical Microwave Antennas

Part 1–Antenna fundamentals and horn antennas

By Paul Wade, N1BWT

(From *QEX*, September 1994*)*

Antenna gain is essential for microwave communication, and since it helps both transmitting and receiving, it is doubly valuable. Practical microwave antennas provide high gain within the range of amateur fabrication skills and budgets.

Three types of microwave antennas meet these criteria: horns, lenses and dishes. Horns are simple, foolproof and easy to build; a 10-GHz horn with 17 dB of gain fits in the palm of a hand. Metal-plate lenses are easy to build, light in weight and noncritical to adjust.[1] Finally, dishes can provide extremely high gain; a 2-foot dish at 10 GHz has more than 30 dB of gain, and much larger dishes are available.

These high gains are only achievable if the antennas are properly implemented. I will try to explain the fundamentals using pictures and graphics as an aid to understanding. In addition, a computer program, *HDL_ANT*, is available for the difficult calculations and details. In this first of three parts, I'll review some basic antenna terminology and concepts and discuss horn antennas. Part 2 will treat dish antennas, and in Part 3 I'll present metal-lens antennas and discuss the microwave antenna measurements needed to verify antenna performance.

Antenna Basics

Before we talk about specific microwave antennas, there are a few common terms that must be defined and explained:

Aperture

The aperture of an antenna is the area that captures energy from a passing radio wave. For a dish antenna, it is not surprising that the aperture is the size of the reflector, and for a horn, the aperture is the area of the mouth of the horn. Wire antennas are not so simple—a thin dipole has almost no area, but its aperture is roughly an ellipse with an area of about $0.13 \lambda^2$. Yagi-Uda antennas have even larger apertures.[2]

Gain

The hypothetical isotropic antenna is a point source that radiates equally in all directions. Any real antenna will radiate

more energy in some directions than in others. Since the antenna cannot create energy, the total power radiated is the same as that of an isotropic antenna driven from the same transmitter; in some directions it radiates more energy than an isotropic antenna, so in others it must radiate less energy. The gain of an antenna in a given direction is the amount of energy radiated in that direction compared to the energy an isotropic antenna would radiate in the same direction when driven with the same input power. Usually we are only interested in the maximum gain—in the direction in which the antenna is radiating most of the power.

An antenna with a large aperture has more gain than a smaller one; just as it captures more energy from a passing radio wave, it also radiates more energy in that direction. Gain may be calculated as:

$$G_{dBi} = 10 \log_{10}\left(\eta \cdot \frac{4\pi}{\lambda^2} \cdot \text{Aperture} \right)$$

with reference to an isotropic radiator; η is the efficiency of the antenna.

Efficiency

Consider a dish antenna pointed at an isotropic antenna transmitting some distance away. We know that the isotropic antenna radiates uniformly in all directions, so it is a simple (!) matter of spherical geometry to calculate how much of that power should be arriving at the dish over its whole aperture. Now we measure how much power is being received from the dish at the electrical connection to the feed—never greater than that arriving at the aperture. The ratio of power received to power arriving is the aperture efficiency.

How much efficiency should we expect? For dishes, all the books say that 55% is reasonable, and 70 to 80% is possible with very good feeds. Several amateur articles have calculated gain based on 65% efficiency, but I haven't found measured data to support any of these numbers. On the other hand, KI4VE suggests that the amateur is lucky to achieve 45-50% efficiency with a small dish and a typical "coffee-can" feed.[3]

For horns and lenses, 50% efficiency is also cited as typical. Thus, we should expect about the same gain from any of these antennas if the aperture area is the same.

---

[1] Notes appear at the end of this section.

Reciprocity

Suppose we transmit alternately with a smaller and a larger dish and note the relative power received at a distant antenna. Then if we transmit from the distant antenna and receive alternately with the same two dishes, would we expect to see the same relative power? Yes. Transmitting and receiving gains and antenna patterns are identical. This is hard to prove mathematically, but it is so.[4,5]

However, the relative noise received by different types of antennas may differ, even with identical antenna gains. Thus, the received signal-to-noise ratio may be better with one type of antenna than another.

Directivity and Beamwidth

Suppose an antenna has 20 dB of gain in some direction. That means it is radiating 100 times as much power in that direction as would an isotropic source, which uniformly distributes its energy over the surface of an arbitrarily large sphere that encloses the antenna. If all the energy from the 20-dB-gain antenna were beamed from the center of that same sphere, it would pass through an area 100 times smaller than the total surface of the sphere. Since there are 41,253 solid degrees in a sphere, the radiation must be concentrated in 1/100th of that, or roughly 20° of beamwidth. The larger the gain, the smaller the beamwidth.

The directivity of an antenna is the maximum gain of the antenna compared to its gain averaged in all directions. It is calculated by calculating the gain, using the previous formula, with 100% efficiency.

Sidelobes

No antenna is able to radiate all the energy in one preferred direction. Some is inevitably radiated in other directions. Often there are small peaks and valleys in the radiated energy as we look in different directions (Fig 1). The peaks are referred to as sidelobes, commonly specified in *dB down from the main lobe*, or preferred direction.

Are sidelobes important? Let's suppose that we could make an antenna with a 1-degree beamwidth, and in all other directions the average radiation was 40 dB down from the main lobe. This seems like a pretty good antenna! Yet when we do the calculation, only 19.5% of the energy is in the main lobe, with the rest in the other 41252/41253 of a sphere. The maximum efficiency this antenna can have is 19.5%.

E-plane and H-plane

An antenna is a transducer which converts voltage and current on a transmission line into an electromagnetic field in space, consisting of an electric field and a magnetic field oriented at right angles to one another. An ordinary dipole creates an electric-field pattern with a larger amplitude in planes which include the dipole than in other planes. The electric field travels in the E-plane; the H-plane, perpendicular to it, is the field in which the magnetic field travels. When we refer to polarization of an antenna, we are referring to the E-plane. However, for three-dimensional antennas like horns, dishes and lenses, it is important to consider both the E-plane and the H-plane, in order to fully use the antenna and achieve maximum gain.

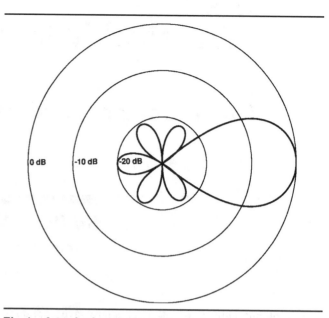

Fig 1—A typical antenna pattern showing the main lobe and sidelobes.

Phase Center

The antenna pattern in Fig 1, and most other illustrations of antenna patterns, shows only amplitude, or average power. This is all we need to consider for most applications, but for antennas which are like optical systems, like lenses and dishes, we must also be concerned with phase, the variation in the signal as a function of time. RF and microwave signals are ac, alternating current, with voltage and current that vary sinusoidally (like waves) with time. Fig 2A shows several sine waves, all at the same frequency, the rate at which they vary with time.

Let's think about a simple example: a child's swing. We've all both ridden and pushed one at some time. If we push the swing just as it starts to move away from us, it swings higher each time. If we add a second pusher at the other end, it will increase faster. Now if we tie a rope to the swing seat and each pusher takes an end, we can try to add energy to the swing throughout its cycle. This will work as long as we keep the pulling synchronized with the motion of the swing, but if we get *out of phase*, we will drag it down rather than sending it higher.

The motion of a swing is periodic, and the height of the swing varies with time in a pattern similar to a sine wave of voltage or current. Look at a sine wave in Fig 2A, considering the highest point of the waveform the height the swing travels forward, and the lowest point as the height the swing travels backward, both repeating with time. If there are two swings side-by-side and both swings arrive at their peak at the same time, they are in phase, as in Fig 2A.

When two electromagnetic waves arrive at a point in space and impinge on an antenna, their relative phase is combined to create a voltage. If they have the same phase, their voltages add together; in Fig 2A, the two dashed waveforms are in phase and add together to form the solid waveform. On the other hand, when signals are exactly out of phase, the addition of positive voltage to negative voltage leaves only the difference, as shown in Fig 2B. If the two signals are partially out of phase, the

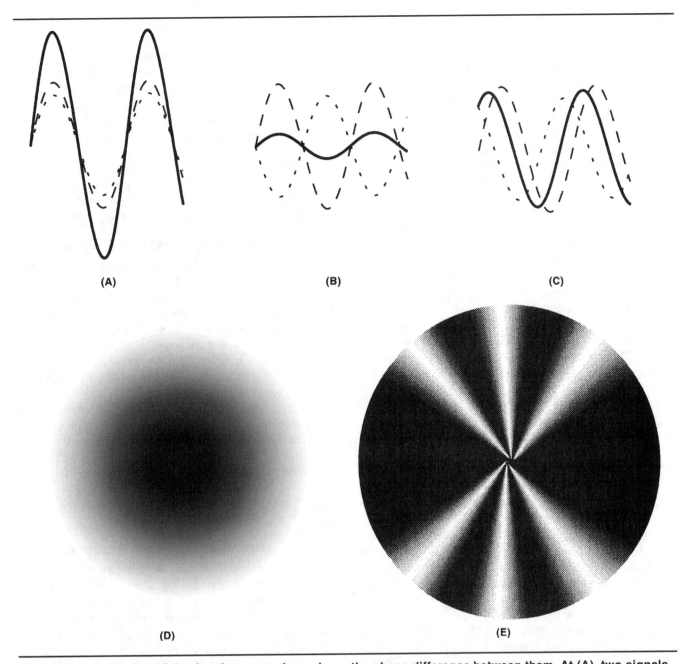

Fig 2—The result of multiple signal sources depends on the phase difference between them. At (A), two signals are shown in phase and add together. At (B), the signals are 180° out of phase and tend to cancel, while the signals at (C) are out of phase by less than 180°, with the result being a signal at a phase and amplitude different from either of the two source signals. The plot at (D) shows the amplitude around a single-source antenna, while (E) shows the interference pattern created by having two sources.

resultant waveform is found by adding the voltage of each at each point in time; one example is shown in Fig 2C. Notice that the amplitude of the resultant waveform is dependent on the phase difference between the two signals.

If our signal source is a point source, then all waves are coming from that one point in space. Each wave has a wavefront, like a wave arriving on a beach. The wavefront from the perfect point source has a spherical shape. Consider its *amplitude*. First, we place an antenna and power meter at some distance from the source and take a reading, then when we move the antenna around to other places that create exactly the same power reading, we will draw a sphere around the source. Thus, the amplitude has a uniform distribution like Fig 2D; dark areas have higher amplitude than lighter areas, and the amplitude decreases as we move away from the source according to the inverse square law described below (the shading has a few small concentric rings due to the limitations of computer graphics, but is really a continuous smooth function).

The *phase* of this wavefront as it propagates in space appears to also have a spherical shape. If frozen in time, one sphere would represent a positive peak of a sine wave. One half wavelength inside would be another sphere representing a negative peak of the sine wave, and another half wave inside again is a positive peak. The *phase center* of an antenna is the apparent place from which the signal emanates based on the center of a sphere of constant phase.

However, no real antenna is small enough to be a point source, so the radiation must appear to emanate from a larger area. If we consider a simple case, where the radiation appears to come from two points, then two signals will arrive at each point in space. A point in space is typically farther from one radiating point than from the other, and since the time it takes for each signal to arrive depends on the distance to each of the radiating points, there will be a phase difference between the two signals. This phase difference will be different at each point in space, depending on the relative distances, and the amplitude of the resultant signal at each point depends on the phase difference. An example of a pattern created by two radiating sources is shown in Fig 2E, where the dark areas have the greatest amplitude, due to the two signals arriving in phase, and the light areas are areas where phase cancellation, like that of Fig 2B, has reduced the amplitude.

A well designed feed for a dish or lens has a single phase center, so the radiation appears to emanate from a single point source. This must be so for at least the main beam, the part of the pattern that illuminates the dish or lens. Away from this main beam, the phase center may move around and appear as multiple points, due to stray reflections and surface currents affecting the radiation pattern. However, since these other directions do not illuminate the dish or lens, they can be ignored.

Inverse Square Law

As two antennas are moved farther apart, received power decreases in proportion to the square of the distance between them; when the distance is doubled, only $1/4$ as much power is received, a reduction of 6 dB. This is because the area illuminated by a given beamwidth angle increases as the square of the distance from the source, so the power per unit area must decrease by the same ratio, the square of the distance. Since the area of the receiving antenna has not changed, the received power must decrease proportionally.

The phase center pattern in Fig 2E does *not* include the effect of inverse square law in the pattern, in order to emphasize the phase cancellation. The effect of including inverse square law would be to lighten the pattern as distance from the phase center increased.

Free Lunch

Since gain is proportional to aperture, larger antennas have more gain than smaller antennas, and poor efficiency can only make a small antenna worse. In spite of various dubious claims by antenna designers and manufacturers, "There's no such thing as a free lunch."[6] All else being equal, the larger the antenna, the greater the gain. But a large antenna with poor efficiency is a waste of metal and money.

Recommended Reading

For those interested in pursuing a deeper understanding of antennas, a number of books are available. A good starting point is *The ARRL Antenna Book* and *The ARRL UHF/Microwave Experimenter's Manual*. [Volume 1 of this book illus-

trates, by way of projects, many antenna concepts.—Ed.] Then there are the classic antenna books, by Kraus, Silver and Jasik.[2,4,7] Lo and Lee have edited a more recent antenna handbook, and Love has compiled most of the significant papers on horns and dishes.[5,8,9] For those interested in computer programming for antenna design, Sletten provides a number of routines.[10] Be warned that the math gets pretty dense once you get beyond the ARRL books.

Summary

This concludes our quick tour through basic antenna concepts and definitions. Now let's apply these concepts to understanding actual microwave antennas, starting with horns.

The *HDL_ANT* Computer Program

The intent of the *HDL_ANT* program is to aid the design of microwave antennas, not to be a whizzy graphics program. The program does the necessary calculations needed to implement a horn, dish or lens antenna, or to design an antenna range and correct the gain measurements. The basic data is entered interactively and results are presented in tabular form. If you like the results, a table of data or a template may be saved to a file for printing or further processing; if not, try another run with new data.

The C++ source code is also included, for those who wish to enhance it or simply to examine the more complex calculations not shown in the text. It has been compiled with Borland C++ version 3.1 and is available from the ARRL BBS at 860-594-0306, or can be downloaded via the Internet from ftp.cs.buffalo.edu in the /pub/ham-radio/qex directory. Alternately, go to the ARRL home page at **http://www.arrl.org** and choose **links**, then **ARRL ftp, QEX** and select *HDL_ANT.ZIP*.

Electromagnetic Horn Antennas

A horn antenna is the ideal choice for a contest rover station. It offers moderate gain in a small, rugged package with no adjustments needed, and has a wide enough beam to be easily pointed under adverse conditions. Fig 3 is a photograph of a homebrew horn mounted on an old Geiger counter case which houses the rest of a 10-GHz wide-band FM transceiver. I have worked six grid squares on 10 GHz from Mt. Wachusett in Massachusetts using a small horn with 17.5 dB of gain.

Horn Design

An antenna may be considered as a transformer from the impedance of a transmission line to the impedance of free space, 377 ohms. A common microwave transmission line is *waveguide*, a hollow pipe carrying an electromagnetic wave.[11] If one dimension of the pipe is greater than a half wavelength, then the wave can propagate through the waveguide with extremely low loss. And if the end of a waveguide is simply left open, the wave will radiate out from the open end.

Practical waveguides have the larger dimension greater than a half wavelength, to allow wave propagation, but smaller than a wavelength, to suppress higher-order *modes* which can interfere with low-loss transmission. Thus the aperture of an open-ended waveguide is less than a wavelength, which does not provide much gain.

For more gain, a larger aperture is desirable, but a larger waveguide is not. However, if the waveguide size is slowly expanded, or tapered, into a larger aperture, then more gain is achieved while preventing undesired modes from reaching the waveguide. This taper is like a funnel, called a conical horn, in cylindrical waveguide. The conical horn for 2304 MHz shown in Fig 4 was made by pop-riveting aluminum flashing to a coffee can. With common rectangular waveguide, the taper creates a familiar pyramidal horn, like those shown in the photograph, Fig 5.

To achieve maximum gain for a given aperture size and maximum efficiency, the taper must be long enough so that the phase of the wave is nearly constant across the aperture. An optimum horn is the shortest one that approaches maximum gain; several definitions are available. The *HDL_ANT* program uses approximate dimensions from a set of tables by Cozzens to design pyramidal horn antennas with gains from 10 to 25 dB.[12] Higher gains are possible, but the length of the horn increases much faster than the gain, so very high gain horns tend to be unwieldy.

Kraus gives the following approximations for beam width in degrees:

$$W_{E-Plane} = \frac{56}{A_{E\lambda}} \quad W_{H-Plane} = \frac{67}{A_{H\lambda}}$$

and dB gain over a dipole:

$$Gain = 10\log_{10}\left(4.5 \cdot A_{E\lambda} \cdot A_{H\lambda}\right)$$

where $A_{E\lambda}$ is the aperture dimension in wavelengths in the E-plane and $A_{H\lambda}$ is the aperture in wavelengths dimension in the H-plane. The *HDL_ANT* program uses a more accurate gain algorithm which corrects the phase error of different taper lengths; for a given aperture, efficiency and gain decrease as the taper is shortened.[13]

Horn Construction

If you are fortunate enough to find a suitable surplus horn, this section is unnecessary. Otherwise, you may want to homebrew one. Horn fabrication is quite simple, so you can homebrew them as needed, for primary antennas with moderate gain or as feeds for higher gain dishes and lenses. Performance of the finished horn almost always matches predictions, with no tuning adjustments required.

The *HDL_ANT* program will design a horn with any desired gain or physical dimensions and then make a template for the horn. The template is a Postscript file; print the file on a computer printer to generate a paper template, tape the paper template to a sheet of copper or brass, cut it out, fold on the dotted lines, and solder the metal horn together on the end of a waveguide. The horn shown in Fig 3 used flashing copper from the local lumberyard, which I soldered together on the kitchen stove.

Fig 6 is a template for a nominal 14-dB horn for 5760 MHz generated by *HDL_ANT*. Try it: copy it on a copier and fold up the copy to see how easy it is to make a horn. It's almost as easy with thin copper. Fig 7 is another template example, a nominal 18-dB horn for 10368 MHz. For horns too large to fit the entire template on one sheet of paper, *HDL_ANT* prints each side on a separate sheet.

Fig 4—A homebrew conical horn for 2304 MHz.

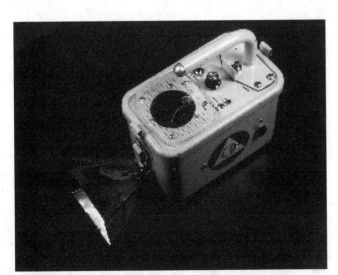

Fig 3—A homebrew horn for 10 GHz, made from flashing copper and designed using the *HDL_ANT* program.

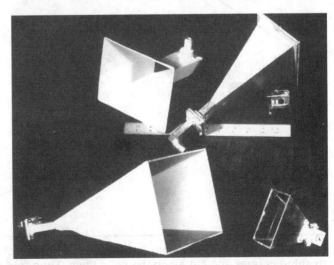

Fig 5—A variety of rectangular horn antennas.

Feed Horns

For horns intended as feed horns for dishes and lenses, beam angle and phase center are more important than horn gain. The *HDL_ANT* program calculates these values in both the E-plane and the H-plane, then allows you to enter new horn dimensions to adjust the beam angle or phase center before making a template. The phase center calculation is a difficult one involving Fresnel sines and cosines, so interactive adjustment of horn dimensions is a lot easier than having the computer try to find the right dimensions.[14,15] The template in Fig 8 is one example of a feed horn—it may be used to make a rectangular horn optimized to feed a dish with f/D = 0.5 at 10 GHz.[16] Feed horn design for dishes and lenses will be described in more detail in the next two sections.

Conclusion

Horns are versatile microwave antennas, easy to design and build with predictable performance. They should be the antenna of choice for all but the highest gain applications.

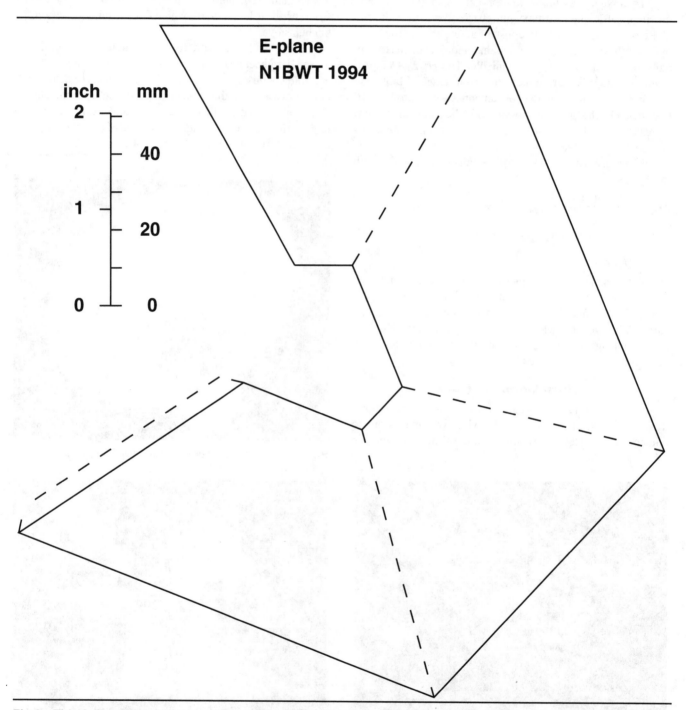

Fig 6—This full-scale template can be used to construct a horn antenna for 5760 MHz. Tape a copy to a piece of flashing copper and cut along the solid lines. Then fold at the dotted lines to form the rectangular horn. Solder the small flap to complete the horn, then solder the narrow end of the horn to a piece of waveguide. This antenna gives 13.8 dBi of gain.

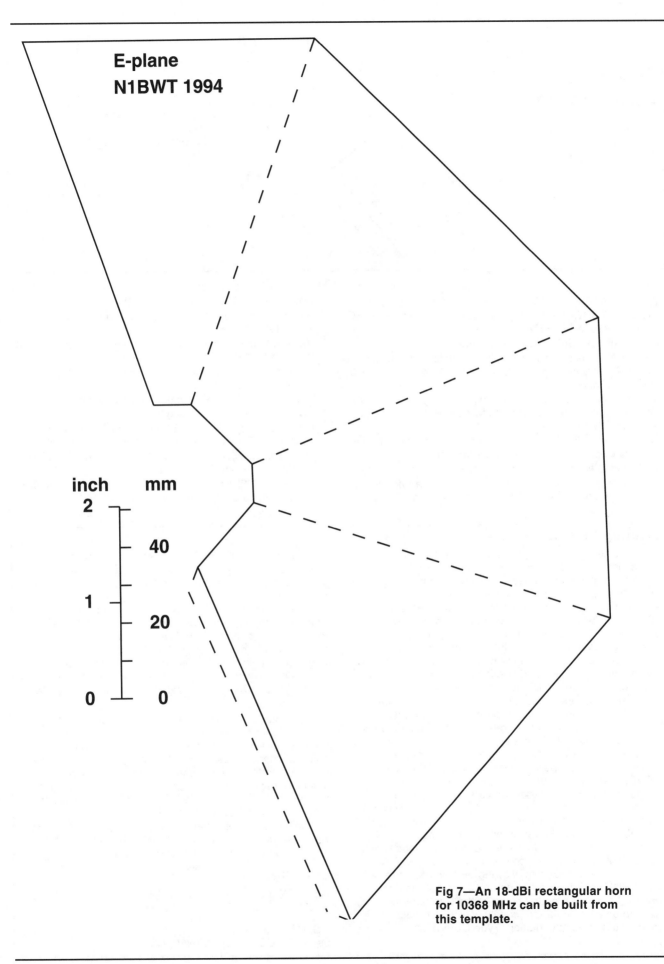

E-plane
N1BWT 1994

inch   mm
2

      40

1

      20

0   0

Fig 7—An 18-dBi rectangular horn for 10368 MHz can be built from this template.

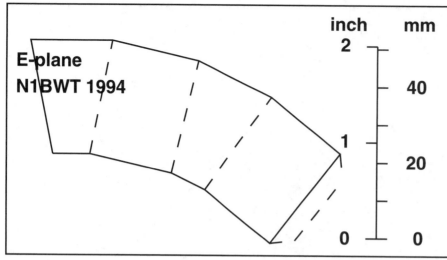

E-plane
N1BWT 1994

inch   mm

2

40

1

20

0   0

Fig 8—A template for a 10368-MHz, 8-dBi horn, suitable for feeding an f/D=0.5 dish.

Notes

[1] Wade, P., N1BWT, and Reilly, M., KB1VC, "Metal Lens Antennas for 10 GHz," *Proceedings of the 18th Eastern VHF/UHF Conference*, ARRL, 1992, pp 71-78.

[2] Kraus, John (W8JK), *Antennas*, McGraw Hill, 1956.

[3] Ralston, M., KI4VE, "Design Considerations for Amateur Microwave Antennas," *Proceedings of Microwave Update '88*, ARRL, 1988, pp 57-59.

[4] Silver, Samuel, *Microwave Antenna Theory and Design*, McGraw-Hill, 1949. (Volume 12 of Radiation Laboratory Series, reprinted 1984.)

[5] Lo, Y. T. and Lee, S. W., editors, *Antenna Handbook: theory, applications, and design*, Van Nostrand Reinhold, 1988.

[6] Attributed to economist Milton Friedman.

[7] Jasik, Henry and Johnson, Richard C., *Antenna Engineering Handbook*, McGraw-Hill, 1984. (Also first edition, 1961.)

[8] Love, A. W., *Electromagnetic Horn Antennas*, IEEE Press, 1976.

[9] Love, A. W., *Reflector Antennas*, IEEE Press, 1978.

[10] Sletten, Carlyle J. (W1YLV), *Reflector and Lens Antennas*, Artech House, 1988.

[11] *The ARRL UHF/Microwave Experimenter's Manual*, ARRL, 1990, pp 5-21 to 5-32.

[12] Cozzens, D. E., "Tables Ease Horn Design," *Microwaves*, March 1966, pp 37-39.

[13] Balanis, C. A., "Horn Antennas," in *Antenna Handbook: theory, applications, and design* (see Note 5).

[14] Muehldorf, Eugen I., *The Phase Center of Horn Antennas*, reprinted in *Electromagnetic Horn Antennas* (see Note 8).

[15] Abramowitz, Milton and Stegun, Irene A., *Handbook of Mathematical Functions*, Dover, 1972.

[16] Evans, D., G3RPE, "Pyramidal horn feeds for paraboloidal dishes," *Radio Communication*, March 1975.

Printing Postscript Files

The easiest way to print Postscript files is with a Postscript compatible laser printer. These have become more affordable and are becoming more common; for instance, the public library in my small town has one attached to a public-access computer. However, they are still roughly twice as expensive as the dot-matrix printers that most of us use with our personal computers.

An alternative to a laser printer is software that interprets Postscript language commands for display on a computer VGA display or a dot-matrix printer. I know of several versions of this type of software. Three commercial products, *GoScript*, *Ultrascript* and *Freedom of Press* perform this function. *Ghostscript*, a freeware program from the Free Software Foundation is available on many bulletin boards and Internet locations. The files are in ZIP format, so they must me downloaded, unZIPped, and installed according to the README documentation.

I have only used *Ghostscript*, version 2.5 and later. The latest versions, *Ghostscript* 5.03 and *Ghostview* 2.3, work very well under Windows 95 and NT; they are available from **http://www.cs.wisc.edu/~ghost/index.html**. They use Unix-style command strings which are difficult to remember, so I've included two BAT files to help: GS_VIEW.BAT for viewing on a screen, and GS_PRINT.BAT for printing on an Epson dot-matrix printer. For other brands of printer, the command will have to be changed appropriately, which will require reading of the documentation. Type GS_VIEW <filename.ps> or GS_PRINT <filename.ps> to use them. Be sure to type QUIT when you are through or your PC may be left in an unhappy state requiring rebooting.

I've included with *HDL_ANT* a sample Postscript file, SQUARE.PS, which draws a four-inch square. Use this to make sure that templates will be drawn to scale. A sample horn template, HORN18.PS, is included, too, to get you started. If the dimensions of the printed square are slightly off, you can correct the scaling. Each template has a line near the beginning of the file:

    1.0 1.0 scale

The first number is the scale factor in the x (horizontal) direction, and the second is the scale factor in the y (vertical) direction. Edit the SQUARE.PS file with an editor to change these numbers slightly; when you find a combination that prints a square exactly four inches on a side, then you have compensated for your printer. Edit these same numbers into any template to be printed on that printer and the dimensions will come out right.

I have not used any of the commercial products, but I would expect a commercial product to be much easier to install and use than freeware or shareware.

Batch Files

GS_VIEW.BAT
gs %1

GS_PRINT.BAT
gs -sDEVICE=epson -r60x60 %1

Practical Microwave Antennas

Part 2–Parabolic Dish Antennas

By Paul Wade, N1BWT

(From *QEX*, October 1994)

Parabolic dish antennas can provide extremely high gains at microwave frequencies. A 2-foot dish at 10 GHz can provide more than 30 dB of gain. The gain is only limited by the size of the parabolic reflector; a number of hams have dishes larger than 20 feet, and occasionally a much larger commercial dish is made available for amateur operation, like the 150-foot one at the Algonquin Radio Observatory in Ontario, used by VE3ONT for the 1993 EME Contest. But these high gains are only achievable if the antennas are properly implemented, and dishes have more critical dimensions than horns and lenses.

Background

Last September (1993), I finished my 10-GHz transverter at 2 PM on the Saturday of the VHF QSO Party. After a quick checkout, I drove up Mt. Wachusett and worked four grids using a small horn antenna. However, for the 10-GHz Contest the following weekend, I wanted to have a better antenna ready.

Several moderate-sized parabolic dish reflectors were available in my garage but lacked feeds and support structures. I had thought this would be no problem, since lots of people, both amateur and commercial, use dish antennas. After reading several articles in the ham literature, I had a fuzzy understanding and was able to put a feed horn on one of the dishes and make a number of contacts of over 200 km from Mt. Washington, in horizontal rain.

But I was not satisfied that I really understood the details of making dishes work, so I got some antenna books from the library and papers from IEEE journals and did some reading. This article is an attempt to explain for others what I've learned. The 10-GHz antenna results from the 1993 Central States VHF Conference suggest that I might not be the only one who is fuzzy on the subject—the dishes measured had efficiencies of from 23% to less than 10%, while all the books say that efficiency should typically be 55%.

[1]Notes appear at the end of this section.

On the other hand, there are enough hams doing successful EME work to suggest that some have mastered feeding their dishes. One of them, VE4MA, has written two good articles on TVRO dishes and feed horns for EME.[1,2]

There have been some good articles written by antenna experts who are also hams, like KI4VE, K5SXK and particularly W2IMU in *The ARRL UHF/Microwave Experimenter's Manual*, which is an excellent starting point. However, as I struggled to understand things that are probably simple and obvious to these folks, I did some reading and then used my personal computer to do some of the more difficult calculations and plot them in ways that helped me to understand what is happening. Many of us find a picture easier to comprehend than a complex equation. What I hope to do here is to start at a very basic level and explain the fundamentals, with pictures and graphics, well enough for hams to implement a dish antenna that works well. An accompanying computer program, *HDL_ANT*, is provided to do the necessary design calculations and to draw templates for small dishes in order to check the accuracy of the parabolic surface. *HDL_ANT* can be downloaded from the ARRL BBS (860-594-0306) or via the Internet from ftp.cs.buffalo.edu in the /pub/ham-radio/qex directory.

Dish Antenna Design

A dish antenna works the same way as a reflecting optical telescope. Electromagnetic waves, either light or radio, arrive on parallel paths from a distant source and are reflected by a mirror to a common point, called the focus. When a ray of light reflects from a mirror or flat surface, the angle of the path it takes leaving the surface (angle of reflection) is the same as the angle at which it arrived (angle of incidence). This optical principle is familiar to anyone who misspent a part of his youth at a pool table! If the mirror is a flat surface, two rays of light that arrive on parallel paths leave on parallel paths; however, if the mirror is curved, two parallel incident rays leave at different angles. If the curve is parabolic, then all the reflected rays meet at one point, as shown

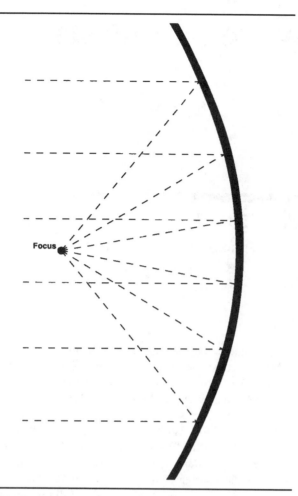

Fig 1—The geometry of a parabolic dish antenna.

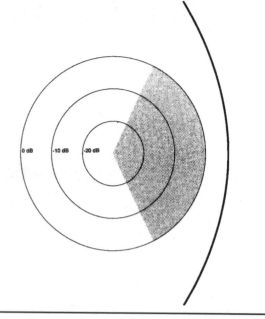

Fig 2—A parabolic dish antenna with uniform feed illumination.

in Fig 1. A dish is a parabola of rotation, a parabolic curve rotated around an axis that passes through the focus and the center of the curve.

A transmitting antenna reverses the path: the light or radio wave originates from a point source at the focus and is reflected into a beam of rays parallel to the axis of the parabola.

Illumination

Some of the difficulties found in real antennas are easier to understand when considering a transmitting antenna but are also present in receiving antennas, since antennas are reciprocal. One difficulty is finding a point source, since any antenna, even a half-wave dipole at 10 GHz, is much bigger than a point. Even if we were able to find a point source, it would radiate equally in all directions, so the energy that was not radiated toward the reflector would be wasted. The energy radiated from the focus toward the reflector illuminates the reflector, just as a light bulb would. So we are looking for a point source that illuminates only the reflector.

Aperture, Gain, and Efficiency

The aperture, gain, and efficiency of an antenna were all defined for antennas in general in Part 1 of this series of articles. The aperture of a dish antenna is the area of the

reflector as seen by a passing radio wave: Aperture = πr^2, where r is the radius, half the diameter of the dish.

If we replace a dish antenna with a much larger one, the greater aperture of the larger dish captures much more of the passing radio wave, so a larger dish has more gain than the smaller one. If we do a little geometry, we find that the gain is proportional to the aperture.

The gain of a dish is calculated as described in Part 1:

$$G_{dBi} = 10 \log\left(\eta \cdot \frac{4\pi}{\lambda^2} \cdot \text{Aperture} \right)$$

with reference to an isotropic radiator, η is the efficiency of the antenna. It might be amusing to calculate the gain of the VE3ONT 150-foot dish at various frequencies; use 50% efficiency to make the first calculation simpler, then try different values to see how efficiency affects gain.

How much efficiency should we expect? All the books say that 55% is reasonable, and 70 to 80% is possible with very good feeds. Several ham articles have calculated gain based on 65% efficiency, but I haven't found measured data to support any of these numbers. On the other hand, KI4VE suggests that the amateur is lucky to achieve 45-50% efficiency with a small dish and a typical "coffee-can" feed.[3]

Practical Dish Antennas

When we first described a parabolic dish antenna, we put a point source at the focus, so that energy would radiate uniformly in all directions both in magnitude and phase. The problem is that the energy that is not radiated toward the reflector will be wasted. What we really want is a feed antenna that radiates only toward the reflector and has a phase pattern that appears to radiate from a single point.

Feed Patterns

We have already seen that efficiency is a measure of

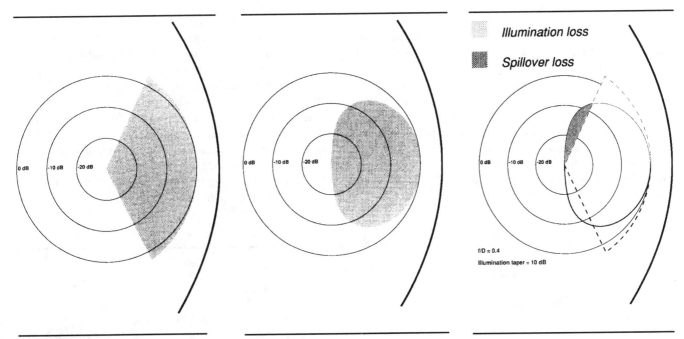

Fig 3—The desired dish illumination would provide uniform field intensity at all points on the reflector.

Fig 4—Typical illumination of a dish using a simple horn feed.

Fig 5—A comparison of typical dish illumination with the desired illuminaton.

how well we use the aperture. If we illuminate the whole reflector, we will be using the whole aperture. Perhaps our feed pattern should be as shown in Fig 2, with uniform feed illumination across the reflector. But when we look more closely at the parabolic surface, we find that the focus is farther from the edge of the reflector than from the center. Since radiated power diminishes with the square of the distance (inverse-square law), less energy is arriving at the edge of the reflector than at the center; this is commonly called space attenuation or space taper. In order to compensate, we must provide more power at the edge of the dish than in the center by adjusting the feed pattern to that shown in Fig 3, in order to have constant illumination over the surface of the reflector.

Simple feed antennas, like a circular horn (coffee-can feed) that many amateurs have used, have a pattern like the idealized pattern shown in Fig 4. In Fig 5 we superimpose that on our desired pattern; we have too much energy in the center, not enough at the edges, and some misses the reflector entirely. The missing energy at the edges is called illumination loss, and the energy that misses the reflector is called spillover loss. The more energy we have at the edge, the more spillover we have, but if we reduce spillover, the outer part of the dish is not well illuminated and is not contributing to the gain. Therefore, simple horns are not ideal dish feeds (although they are useful). In order to have very efficient dish illumination we need to increase energy near the edge of the dish and have the energy drop off very quickly beyond the edge.

Edge Taper

Almost all feed horns will provide less energy at the edge of the dish than at the center, like Fig 4. The difference in power at the edge is referred to as the *edge taper*, or *illumination taper*. With different feed horns, we can vary the edge taper with which a dish is illuminated. Different edge tapers produce different amounts of illumination loss and spillover loss, as shown in Fig 6: a small edge taper results in larger spillover loss, while a large edge taper reduces the spillover loss at the expense of increased illumination loss.

If we plot these losses versus the energy at the edge of the dish in Fig 7, we find that the total efficiency of a dish antenna peaks with an illumination taper, like Fig 6, so that the energy at the edge is about 10 dB lower than the energy at the center.[4,6] This is often referred to as 10-dB edge taper or edge illumination—often recommended but not explained.

G/T

When an antenna is receiving a signal from space, such as a satellite or EME signal, there is very little background noise emanating from the sky compared to the noise generated by the warm (300 K) Earth during terrestrial communications. Most of the noise received by an antenna pointed at the sky is earth noise arriving through feed spillover. As we saw in Fig 6, the spillover can be reduced by increasing the edge taper, while Fig 7 shows the efficiency, and thus the gain, decreasing slowly as edge taper is increased. The best compromise is reached when *G/T*, the ratio of gain to antenna noise temperature, is maximum. This typically occurs with an edge taper of about 13 dB, but the optimum edge taper for *G/T* is a function of receiver noise temperature and sky noise temperature at any given frequency.[2]

Focal Length and *f/D* Ratio

All parabolic dishes have a parabolic curvature, but some are shallow dishes, while others are much deeper and more like a bowl. They are just different parts of a parabola that extends to infinity. A convenient way to describe how much of the parabola is used is the *f/D* ratio, the ratio of the focal length *f* to the diameter *D* of the dish. All dishes with the same *f/D* ratio require the same feed geometry, in proportion to the diameter of the dish. The figures so far have depicted one arbitrary *f/D*; Fig 8 shows the relative geometries for commonly used *f/D* ratios, from 0.25 to 0.65, with the desired and idealized feed patterns for each.

Notice the feed horn patterns for the various *f/D* ratios in Fig 8. As *f/D* becomes smaller, the feed pattern to illuminate it becomes broader, so different feed horns are needed to properly illuminate dishes with different *f/D* ratios. The feed horn pattern must be matched to the reflector *f/D*. Larger *f/D* dishes need a feed horn with a moderate beamwidth, while a dish with an *f/D* of 0.25 has the focus level with the edge of the dish, so the subtended angle that must be illuminated is 180 degrees. Also, the edge of the dish is twice as far from the focus as the center of the dish, so the desired pattern would have to be 6 dB stronger (inverse-square law) at the edge as in the center.

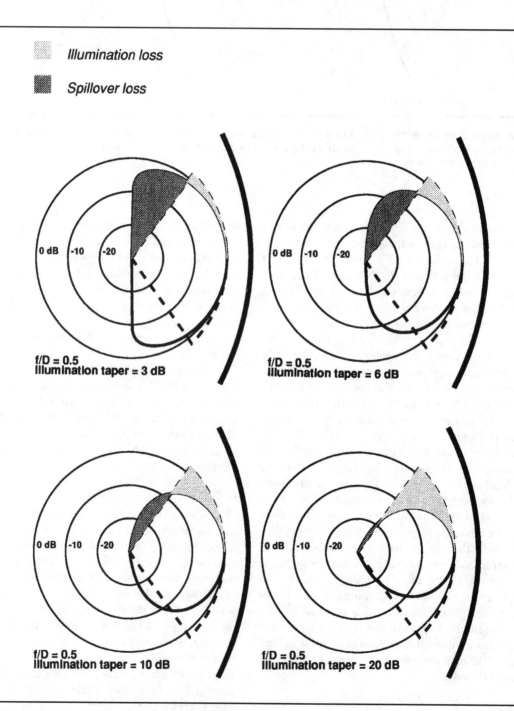

Fig 6—Dish illumination at various values of illumination taper.

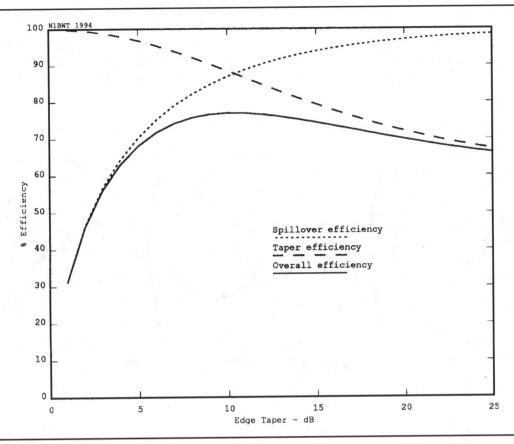

Fig 7—Dish efficiency versus edge taper. The peak efficiency occurs at a taper of about 10 dB.

This is an extremely difficult feed pattern to generate. Consequently, it is almost impossible to efficiently illuminate a dish this deep.

Phase Center

A well-designed feed for a dish or lens has a single phase center, as described in Part 1 of this series of articles, so that the feed radiation appears to emanate from a single point source, at least for the main beam, the part of the pattern that illuminates the dish or lens. Away from the main beam, the phase center may move around and appear as multiple points, as stray reflections and surface currents affect the radiation pattern. Also, the phase center will move with frequency, adding difficulty to broadband feed design. Fortunately, we are only considering narrow frequency ranges here.

Symmetry of E-Plane and H-Plane

On paper, we can only depict radiation in one plane. For a simple antenna with linear polarization, like a dipole, this is all we really care about. A dish, however, is three-dimensional, so we must feed it uniformly in all planes. The usual plane for linear polarization is the E-plane, while the plane perpendicular to it is the H-plane. Unfortunately, most antennas not only have different radiation patterns in the E- and H-planes, but also have different phase centers in each plane, so both phase centers cannot be at the focus.

Table 1
Measured Effect of Focal Length Error at 10 GHz

Feed Distance (in)	Relative Gain (dB)
8.125	−0.6
8.25	0
8.375	−0.3
8.625	−1.7

Focal Length Error

When I started actually measuring the gain of dish antennas, I discovered the most critical dimension to be the focal length—the axial distance from the feed to the center of the dish. A change of ¼ inch, or about a quarter-wavelength, changed the gain by a dB or more, shown in Table 1 as measured on a 22-inch dish with $f/D = 0.39$.

I was surprised at this sensitivity, since my experience with optics and photography suggested that this is not so critical—it would be extremely difficult to adjust a lens or telescope to an optical quarter wavelength. But lenses become more critical to focus as the f-stop is decreased—an f 2 lens is considered to have a very small depth of field, while an f 16 lens has a large depth of field, or broad focus. The f-stop of a lens is the same as the f/D ratio of a dish—both are the ratio of the focal length to the aperture diameter.

A typical reflector telescope has a parabolic reflector of $f8$, but a dish antenna with $f/D = 0.4$ has an f-stop of 0.4, so focusing is much more critical.

More reading located an article which described how to calculate the loss due to focal length error.[7] Fig 9 shows the loss as the feed horn is moved closer and farther than the focus for various f/D dishes with uniform illumination; the tapered illumination we use in practice will not have nulls as deep as the curves shown in Fig 9. It is clear that dishes with small f/D are much more sensitive to focal length error.

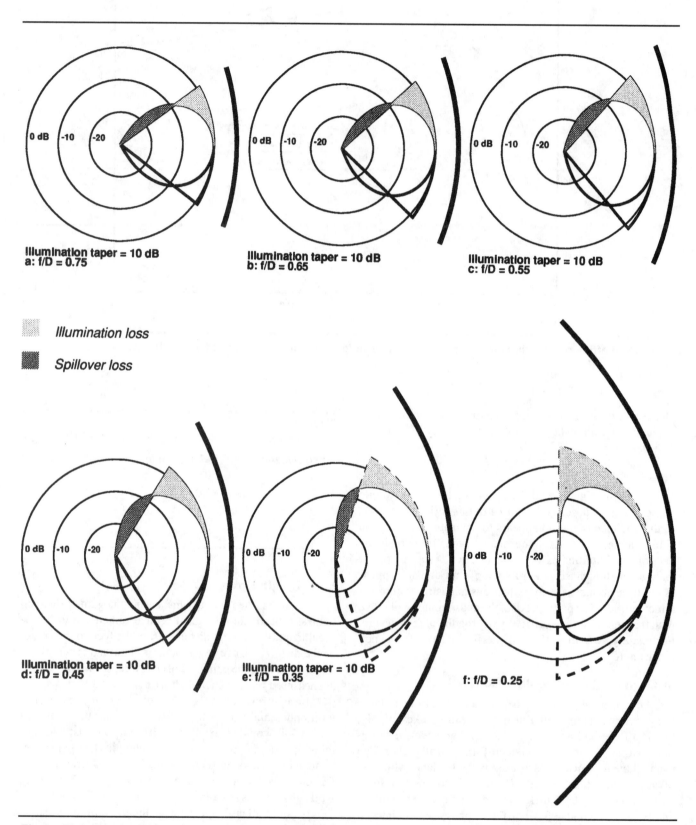

Fig 8—Dish illumination patterns for dishes of various **f/D** ratios.

Remember that a wavelength at 10 GHz is just over an inch.

The critical focal length suggests that it is crucial to have the phase center of the feed exactly at the focus of the reflector. Since the phase center is rarely specified for a feed horn, we must determine it empirically, by finding the maximum gain on a reflector with known focal length.

If we are using a feed horn with different phase centers in the E- and H-planes, we can also estimate the loss suffered in each plane by referring to Fig 9.

Lateral errors in feed horn position are far less serious;

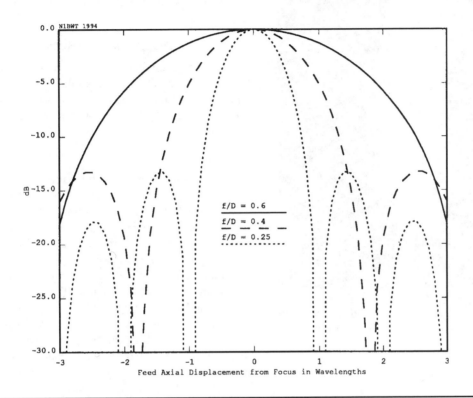

Fig 9—The loss due to axial displacement of the feed from the focus point is highly dependent on the *f/D* ratio.

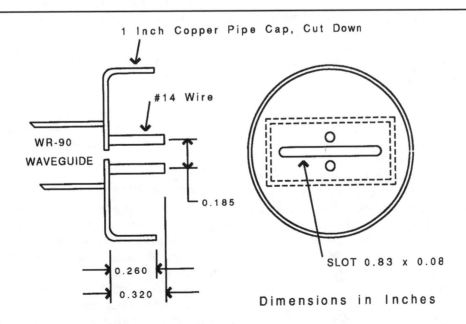

Fig 10—A Clavin feed for 10 GHz, made from a 1-inch copper pipe cap.

Fig 11—This photo shows the technique of mounting a dish using a frying pan with a rolled edge. Also note the Clavin feed used with this dish.

small errors have little effect on gain, but do result in shifting the beam slightly off bore-sight.

Notice that the focal-length error in Fig 9 is in *wavelengths*, independent of the dish size. A quarter-wavelength error in focal length produces the same loss for a 150-foot dish as for a 2-foot dish, and a quarter-wavelength at 10-GHz is just over 1/4 inch. Another implication is that multiband feeds, like the WA3RMX triband feed, should be optimized for the highest band, since they will be less critical at lower bands with longer wavelengths.[8]

Total Efficiency

It has been fairly easy to calculate efficiency for an idealized feed horn pattern due to illumination taper and spillover, but there are several other factors that can significantly reduce efficiency. Because the feed horn and its supporting structures are in the beam of the dish, part of the radiation is blocked or deflected. A real feed horn also has sidelobes, so part of its radiation is in undesired directions and thus wasted. Finally, no reflector is a perfect parabola, so the focusing of the beam is not perfect. We end up with quite a list of contributions to total efficiency:

- illumination taper
- spillover loss
- asymmetries in the E- and H-planes
- focal point error

- feed horn sidelobes
- blockage by the feed horn
- blockage by supporting structures
- imperfections in parabolic surface
- feed line loss

KI4VE suggests that the amateur is lucky to achieve 45-50% efficiency with a small dish and a typical coffee-can feed.[3] I suspect that the only way to find total efficiency, or to optimize it, is to make gain measurements on the complete antenna.

Practical Feed Systems

An optimum feed would approximate the desired feed pattern for the *f/D* of the parabolic reflector in both planes and have the same phase center in both planes. Let's examine some of the available feed horn designs to see how well they do:

1. Dipole

Most hams know what the pattern from a dipole looks like—in free space, it looks like a donut with the dipole through the hole. If it is near ground or a reflector, the pattern in the plane perpendicular to the dipole (H-plane) is distorted to emphasize radiation away from the reflector. The shape of the radiation in this plane is controlled by the distance from the reflector, while the shape of the radiation in a plane parallel to the dipole (E-plane) does not change significantly. This suggests that the best we could do is to find a dish with an edge angle that approximates the E-plane beamwidth and adjust the reflector spacing so that the H-plane beamwidth matches the E-plane. Round disc reflectors are frequently used, but it turns out that the pattern is the same as a half-wavelength rod reflector.

2. Dual Dipole

The H-plane beamwidth can be narrowed by adding a second parallel dipole over a plane reflector, such as the EIA (sometimes erroneously called NBS) reference antenna.[9] This is a reasonably good feed with good symmetry for reflectors with *f/D* around 0.55 and has been used with good success for 432-MHz EME.

3. Penny-Splasher

The penny-splasher feed is equivalent to a dual dipole with reflector—the slots in the waveguide act as dipoles.[10] In practice, however, it has poor sidelobes that result in low efficiency.

4. Rectangular Horn

The beamwidth of a horn antenna is controlled by the horn aperture dimension, but a square horn has different E- and H-plane beamwidths. We can make it rectangular with the aperture dimensions adjusted so that the E- and H-plane patterns and beamwidths are similar. G3RPE described this technique and showed that at 10 GHz it can only illuminate dishes with *f/D* greater than 0.48 if the horn is driven by common WR-90 waveguide.[11] However, the smaller WR-75 waveguide is also suitable for 10 GHz and could drive horns which would illuminate an *f/D* as small as 0.43.

With a rectangular horn, it is difficult to achieve both a common phase center for the E- and H-planes and similar patterns in both planes. The horn section of the *HDL_ANT* computer program calculates the phase centers and allows adjustment of dimensions to change them. Kraus shows a series of patterns for horns with different flare angles, and some of them approximate the desirable feed pattern of Fig 3.[12] However, no phase information is given; W2IMU once told me they were terrible, and I accept his authority.

5. Circular Horn

A circular horn antenna, since it is symmetrical, might be expected to provide a fairly symmetrical pattern. Unfortunately, it doesn't, and the phase centers are different for the E- and H-planes. The beamwidth is controlled by the diameter of the horn—for wide beamwidths, the horn may have no flare, like the coffee-can feed, or cylindrical horn, often used at 1296 MHz.[13]

Some improvement in the pattern may be provided by adding a choke flange to a cylindrical horn.[14] Further improvement is possible by adding slots in the flange, though radiation patterns are shown in only one plane.[15]

All of the above feeds have patterns similar to Fig 4. Many of these were developed for radar applications, where feed inefficiency may be compensated by increased power. More recently, satellite communication has prompted research into more efficient feed antennas, particularly for deep dishes (small f/D) with reduced sidelobes and better G/T. Here are a few of the many variations that have been described, chosen for their potential for construction without elaborate machining:

6. Clavin Feed

The Clavin feed is a cavity antenna fed by a resonant slot, with probes that excite a second waveguide mode to broaden the pattern in the H-plane to match the E-plane.[16] Radiation patterns approximate our desired feed pattern, Fig 3, while maintaining a good phase center. Fig 10 is a sketch of one I made from a 1-inch copper plumbing pipe cap. It is best for deep dishes with *f/D* in the 0.35 to 0.4 range. The resonant slot makes it more narrowband than the others (not a problem for amateur use), and the smaller size would have less feed blockage than the "Chaparral" or Kumar feeds, so it might provide better performance on smaller dishes.

A scalar feed is one that has no inherent polarization; the word "scalar" means that the electric field distribution is independent of the axis in which you look at the distribution. The result is that scalar horns have equal beamwidths and sidelobes in both azimuth and elevation. This can't be achieved with a standard flared horn, so scalar horns are usually preferred for dish feeds. The symmetry also makes them suitable for both linear and circular polarization. The W2IMU dual-mode horn and the "Chaparral" and Kumar feeds below are scalar feeds.

7. W2IMU Dual-Mode Horn

Diffraction from the edge of a horn causes sidelobes that reduce efficiency. In the W2IMU dual-mode horn design, there is a flare from a small section, which only supports the lowest waveguide mode, to a larger section that supports two waveguide modes.[5,17,18] The size of the flare controls the relative amplitude of the two modes, and the length of the large section is chosen so that the two modes cancel at the edge of the horn because they travel at different phase velocities in the waveguide. The cancellation eliminates the sidelobes and thus puts more energy onto the reflector. The requirement for a larger horn makes this feed optimum for larger *f/D* reflectors, in the 0.5 to 0.6 range.

8. Chaparral Feed

The "Chaparral" feed is a type of scalar feed horn often found on TVRO dishes, with a series of cavity rings surrounding a circular waveguide.[19,5,4] The rings modify the pattern to approximate our desired feed pattern, Fig 3, while maintaining a good phase center. This feed is best for deep dishes, with *f/D* in the 0.35 to 0.45 range. Fine adjustment of the pattern is possible by changing the protrusion of the central waveguide in relation to the surrounding rings.

Note: I have not seen any mention of the location of the phase center, but my experiments show that it is controlled by the location of the outer rings, not the central waveguide.

9. Kumar Feed

The Kumar feed is a scalar feed horn similar to the Chaparral feed, but with a single larger outer ring, so construction is somewhat simpler.[20] Radiation patterns approximate our desired feed pattern, Fig 3, while maintaining a good phase center. Ham-band versions of this feed have been described by VE4MA for 1296, 2304 and 3456 MHz.[2,21] Like the Chaparral feed, it is best for deep dishes, with *f/D* in the 0.35 to 0.45 range, with similar fine adjustment.

Complete Dish Antennas

Many of the papers describing feed horns show great detail of the horn performance, but very few even mention what happens when a reflector is added. The reflector may add too many uncertainties for good research, but our goal is to make a good working antenna. We want high efficiency because a dish has the same size, wind loading, and narrow beamwidth regardless of efficiency—we should get as much performance as possible for these operational difficulties. In other words, if I am going to struggle with a one-meter diameter dish on a windy mountain top, I certainly want one meter worth of performance!

In order to compare the different feeds, I wanted to measure the gain of several of them with the same reflector, to find their performance as complete antennas. I made a mechanism from an old slotted-line carriage and some photographic hardware that allows the feed to be moved in three dimensions with fine control of adjustment, so the feed position can be adjusted for maximum gain.

The emphasis here is on smaller dishes intended for mountaintopping and other portable operation, so maximum gain with minimum size and weight is a definite consideration. For other applications, there would be other considerations; EME, for instance, would mandate maximum performance.

Parabolic Reflector

I have managed to collect a half-dozen parabolic reflectors of various sizes and origins, and I wanted to know if they were useful at 10 GHz. First, for each dish I measured the diameter and depth in the center of the dish in order to calculate the focal length and f/D ratio . This can only be an approximation for some dishes, due to holes or flat areas in the center. The focal length is calculated as:

$$f = \frac{D^2}{16 \cdot depth}$$

The *HDL_ANT* computer program does the calculation and then generates a Postscript plot of a parabolic curve for the specified diameter and f/D ratio. For each reflector, I made a series of plots on a laser printer for a range of f/D values for antennas in general near the calculated value, cut out templates, then fitted them to the surface to find the closest fit. For 10 GHz, the surface must be within ±1 mm of a true parabola for optimum performance, although errors up to ±3 mm result in only 1 dB degradation.[22] I selected several reflectors with good surfaces and discarded one that wasn't even close.

Given a choice, a reflector with a large f/D (0.5 to 0.6) would be preferable. As described earlier, dishes with small f/D are hard to illuminate efficiently and are more sensitive to focal length errors. On the other hand, a dish that is available for the right price is always a good starting point!

Parabolic reflectors can come from many sources, not just antenna manufacturers. Some aluminum snow coasters (now unfortunately replaced by plastic, but aluminum foil glued to the surface might make them usable) are good, and hams in Great Britain have put dustbin lids into service as effective parabolic reflectors for years.

Homebrewing a parabolic reflector is possible, but great difficulty is implied by the surface accuracy cited above. The surface accuracy requirement scales with wavelength, so the task is easier at lower frequencies. Of course, hams are always resourceful—N1IOL found that the cover from his 100-pound propane tank was an excellent 14-inch parabolic surface and has used it to mold a number of fiberglass reflectors. K1LPS then borrowed a larger cover from a different type of propane tank and found it to be nowhere near a parabola!

Recommended Feed Systems

Since no single feed system is optimum for all dishes, a good feed recommendation depends on the f/D of the particular dish. For shallow dishes (f/D of 0.5 to 0.6), I'd recommend the W2IMU dual-mode horn or a pyramidal horn designed for the exact f/D.[5,11] The horn section of the computer program will design the horn and plot a construction template. For deeper dishes (f/D of 0.3 to 0.45), I'd recommend the Chaparral, Kumar or Clavin feeds.[5,20,16] For 10 GHz, a Chaparral horn designed for 11-GHz TVRO use works well; your local satellite TV dealer might be persuaded to order it as an "11 GHz Superfeed."

Mechanical Support

There are two critical mechanical problems: mounting the feed horn to the dish and mounting the dish to the tripod. Most small dishes have no backing structure, so the thin aluminum surface is easily deformed. K1LPS discovered that some cast-aluminum frying pans have a rolled edge that sits nicely on the back of a dish; Mirro is one suitable brand. This is a good use for that old frying pan with the worn-out Teflon coating, so buy a new one for the kitchen. Tap a few holes in the edge of the old pan, screw the dish to it, and you have a solid backing. A solid piece of angle iron or aluminum attaches the bottom of the frying pan to the top of a tripod. The photograph in Fig 11 shows a dish mounted using a frying pan. WA1MBA uses this technique for a 24-inch dish at his home and reports that it stands up well to New England winters.

The mounting structure for the feed horn is in the RF field, so we must minimize the blockage it causes. We do this by using insulating materials and by mounting the support struts diagonally, so they aren't in the plane of the polarization. Fiberglass is a good material; plant stakes or bicycle flags are good sources, and WA5VJB recommends cheap target arrows. Use of four rather than three struts is recommended—if they are all the same length, then the feed is centered. The base of the struts should be attached to the backing structure or edge of the frying pan; the thin dish surface is not mechanically strong.

Aiming

A quality compass and a way of accurately aligning the antenna to it are essential for successful operation. Narrow beamwidth and frequency uncertainty can make searching for weak signals frustrating and time-consuming. A heavy tripod with setting circles is a good start; hang your battery from the center of the tripod and it won't blow over as often. Calibrate your headings by locating a station with a known beam heading rather than by eyeballing the dish heading; small mechanical tolerances can easily shift the beam a few degrees from the apparent boresight. As W1AIM can testify, having the wind blow a dish over can distort it enough to move the beam to an entirely different heading.

Alternatives

The narrow beamwidth may actually make contacts more difficult, particularly in windy conditions. I have worked six grids from Mt. Wachusett in central Massachusetts using a small Gunnplexer horn. The longest path, 203 km, required a 12-inch lens for additional gain to make the contact on wideband FM; it would have been easy with narrowband SSB or CW.[23]

For a rover station, a reasonable size horn might be a good compromise, with adequate gain and moderate beamwidth for easy aiming. I often use the 17.5-dBi Gunnplexer horn, with a 12-inch lens ready to place in front of it when signals are marginal.

Conclusions

A parabolic dish antenna can provide very high gain at microwave frequencies, but only with very sharp beamwidths. To achieve optimum gain, careful attention to detail is required: checking the parabolic surface accuracy with a template, matching the feed horn to the *f/D* of the dish, and, most importantly, accurately locating the phase center of the feed horn at the focus.

Notes

[1]Malowanchuk, B. W., VE4MA, "Use of Small TVRO Dishes for EME," *Proceedings of the 21st Conference of the Central States VHF Society*, ARRL, 1987, pp 68-77.

[2]Malowanchuk, B. W., VE4MA, "Selection of an Optimum Dish Feed," *Proceedings of the 23rd Conference of the Central States VHF Society*, ARRL, 1989, pp 35-43.

[3]Ralston, M., KI4VE, "Design Considerations for Amateur Microwave Antennas," *Proceedings of Microwave Update '88*, ARRL, 1988, pp 57-59.

[4]Reasoner, H., K5SXK, "Microwave Feeds for Parabolic Dishes," *Proceedings of Microwave Update '89*, ARRL, 1989, pp 75-84.

[5]Turrin, D., W2IMU, "Parabolic Reflector Antennas and Feeds," *The ARRL UHF/Microwave Experimenter's Manual*, ARRL, 1990.

[6]Rahmat-Samii, Y., "Reflector Antennas," *Antenna Handbook: theory, applications, and design*, Y.T. Lo and S.W. Lee, editors, Van Nostrand Reinhold, 1988, p 15-42.

[7]Ingerson, P. G. and Rusch, W. V. T., "Radiation from a paraboloid with an axially defocused feed," *IEEE Transactions on Antennas and Propagation*, Vol AP-21 No. 1, Jan 1973, pp 104-106.

[8]Hill, T., WA3RMX, "A Triband Microwave Dish Feed," *QST*, Aug 1990, pp 23-27.

[9]Turrin, D., W2IMU, "Antenna Performance Measurements," *QST*, Nov 1974, pp 35-41.

[10]Heidemann, R., DC3QS, "A Simple Radiator for 3 cm Parabolic Dishes," *VHF Communications*, 3/1979, pp 151-153.

[11]Evans, D., G3RPE, "Pyramidal horn feeds for paraboloidal dishes," *Radio Communication*, March 1975.

[12]Kraus, John, W8JK, *Antennas*, McGraw Hill, 1956.

[13]Vilardi, D., WA2VTR, "Simple and Efficient Feed for Parabolic Antennas," *QST*, March 1973, pp 43-44.

[14]Foot, N. J., WA9HUV, "Second-generation cylindrical feedhorns," *Ham Radio*, May 1982, pp 31-35.

[15]Foot, N. J., WA9HUV, "Cylindrical feedhorns revisited," *Ham Radio*, Feb 1986, pp 20-22.

[16]Clavin, A., "A Multimode Antenna Having Equal E- and H-Planes," *IEEE Transactions on Antennas and Propagation*, AP-23, Sep 1975, pp 735-737.

[17]Turrin, R. H., W2IMU, "Dual Mode Small-Aperture Antennas," *IEEE Transactions on Antennas and Propagation*, AP15, Mar 1967, pp 307-308.

[18]Turrin, D., W2IMU, "A Simple Dual-Mode (IMU) Feed Antenna for 10368 MHz," *Proceedings of Microwave Update '91*, ARRL, 1991, p 309.

[19]Wohlleben, R., Mattes, H. and Lochner, O., "Simple small primary feed for large opening angles and high aperture efficiency," *Electronics Letters*, 21 Sep 1972, pp 181-183.

[20]Kumar, A., "Reduce Cross-Polarization in Reflector-Type Antennas," *Microwaves*, Mar 1978, pp 48-51.

[21]Malowanchuk, B. W., VE4MA, "VE4MA 3456 MHz circular polarization feed horn," North Texas Microwave Society *Feedpoint*, Nov/Dec 1991.

[22]Knadle, R. T., K2RIW, "A Twelve-Foot Stressed Parabolic Dish," *QST*, Aug 1972, pp 16-22.

[23]Wade, P., N1BWT, and Reilly, M., KB1VC, "Metal Lens Antennas for 10 GHz," *Proceedings of the 18th Eastern VHF/UHF Conference*, ARRL, May 1992, pp 71-78.

Practical Microwave Antennas

Part 3–Lens antennas and microwave antenna measurements.

By Paul Wade, N1BWT

(From *QEX*, November 1994)

In the previous parts of this series we discussed horn antennas and parabolic dish antennas. We now turn our attention to the third type of practical microwave antennas: lenses. I'll also describe microwave antenna measurement techniques and conclude with a discussion of actual measurement results and a comparison of the three types of antennas.

Lens Antennas

For portable microwave operation, particularly if backpacking is necessary, dishes or large horns may be heavy and bulky to carry. A metal-plate lens antenna is an attractive alternative. Placed in front of a modest-sized horn, the lens provides some additional gain, much like eyeglasses on a near-sighted person. The lens antennas I have built and tested are cheap and easy to construct, light in weight and noncritical to adjust. The *HDL_ANT* computer program makes designing them easy, as well.

There are other forms of microwave lenses—for instance, dielectric lenses and Fresnel lenses—but the metal-plate lens is probably the easiest to build and lightest to carry, so it is the only type I'll describe here.

The metal-lens antenna is constructed of a series of thin metal plates with air between them. The curvature of the edges of the plates forms the lens, and the space between the plates forms a series of waveguides. Fortunately, we can get·"air" in a solid form to make construction easier: Styrofoam looks just like air to RF, and it keeps the metal plates accurately spaced. We use aluminum foil for the plates, attaching it to the Styrofoam with spray adhesive and shaping the curvature with a hobby knife on a compass. Designs are limited to those using circles, to ease construction.

Background

These metal-plate lenses were originally described for 10 GHz by KB1VC and me at the 1992 Eastern VHF/UHF Conference, but there is no good reason to limit them to that band.[1] The need for more gain became apparent to us during

[1]Notes appear at the end of this section.

the 1991 10-GHz Contest. We were atop Burke Mountain in Vermont, on a day as clear as the tourist brochures promise. We could see Mt. Greylock in Massachusetts, where KH6CP was located, but it was too far to work with horn antennas on our Gunnplexers. After K1LPS humped his two-foot dish up the fire tower, we knew that wasn't the best answer for portable work.

Later, we found an article in *VHF Communications* on lens antennas by Angel Vilaseca, HB9SLV, which intrigued us.[2] It described how to design a metal-lens antenna but did not present expected gain or measured results.

We then searched through the references to try to understand how these antennas work, finally discovering that the best work was done before we were born, by Kock.[3] Kock's paper makes it clear how the metal-lens antenna works, and, more importantly, that it *does* work!

Lens Basics

The metal-plate lens works, in principle, like any other lens. A similar optical lens would take a broad beam of light and shape it, by refraction, into a narrower beam.[4] Refraction occurs at the interface of two materials in which light travels at different speeds and changes the direction of travel of the beam of light. If the beam is formed of many rays of light, each one may be bent; the ones at the edge of the beam bend more so they end up parallel to the center rays, which are hardly bent at all. For this to work, each ray must take exactly the same time to travel from its source, at the focal point of the lens, to its destination. Since light travels more slowly in glass, a lens is thicker at the middle, to slow down the rays with a shorter path, and thinner at the edges, to allow the rays with longer paths to catch up, as shown in Fig 1. The needed curvature of the lens to form the beam exactly is an ellipse, but for small bending angles a circle is almost identical to an ellipse, and nearly all optical lenses are ground with spherical curves.

Since light and RF are both electromagnetic waves, we could use glass—or any other dielectric—to make a lens for 10 GHz. For example, a recent article described a dielectric lens made of epoxy resin.[5] But for larger sizes this quickly

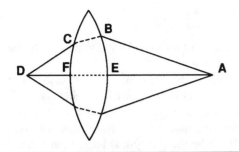

Fig 1—A simple lens. The travel time for each of the rays must be the same, so the time along the line ABCD is the same as that along the line AEFD.

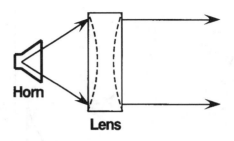

Fig 2—Feeding a lens with a horn lets the horn provide part of the beam shaping.

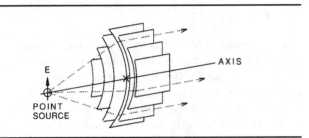

Fig 3—A spherical lens can be formed by a series of spaced plates.

becomes less attractive, and most dielectrics are rather lossy at 10 GHz. Low-loss materials are available but are costly and relatively heavy and difficult to shape.

The Metal-Plate Lens

Since electromagnetic waves travel at different speeds in waveguide and in free space, why not use waveguides of different lengths to form a lens? This has been done and is known as an "eggcrate" lens.[6] However, it is easier to make a group of parallel plates that form wide parallel waveguides, simply shaping the input and output edges of these waveguides to change the path lengths and form the lens surface. This differs from an optical lens in that the phase of the electromagnetic wave travels *faster* in a waveguide than in free space.[7] Thus, the required curvature of a metal lens antenna is the *opposite* of an equivalent optical or dielectric lens—in this case, concave instead of convex. We can still get away with using circular curvatures instead of ellipses as long as we aren't trying to bend the rays too sharply. For that reason, we feed the lens with a small horn, which does part of the beam forming, as shown in Fig 2. Of course, if we want both horizontal and vertical beam shaping, we need a spherical shape, so we must shape the surface described by the edges of the metal plates into a sphere like that of Fig 3.

Lens Design

While the *HDL_ANT* program removes the drudgery from lens design and makes it available to amateurs, a gen-

eral description of lens design might aid in understanding what is happening and what the computer is telling you.

First, some design objectives are needed: how big a lens is desired, and what are the dimensions of the horn feeding it? Gain is determined by aperture (roughly the diameter for dishes, horns and lenses). A good rule of thumb is that doubling the aperture will increase the gain by 6 dB. For instance, an 8-inch lens in front of a 4-inch horn would add 6 dB to the gain of the horn, and a 16-inch lens would add 12 dB. So, modest gain improvements take modest sizes, but really large gains require huge antennas no matter what kind. However, a 6-dB increase in gain will double the range of a system over a line-of-sight path.

The horn dimensions may be determined by availability, or you may have the design freedom to build the horn as well. The beam width of the horn (which is usually smaller than the physical flare angle of the horn) is used to determine the focal length of the lens. Kraus gives the following approximations for beam width in degrees and dB gain over a dipole:[8]

$$W_{Eplane} = \frac{56}{A_{E\lambda}} \qquad (Eq\ 1)$$

$$W_{Hplane} = \frac{67}{A_{H\lambda}} \qquad (Eq\ 2)$$

$$Gain = 10\log_{10}\left(4.5A_{E\lambda}A_{H\lambda}\right) \qquad (Eq\ 3)$$

where $A_{E\lambda}$ is the aperture dimension in wavelengths in the E-plane, and $A_{H\lambda}$ is the aperture in wavelengths in the H-plane. These approximations are accurate enough to begin designing. From the beam width and desired lens aperture, finding the focal length f is a matter of geometry:

$$f = \frac{Lens\ diameter}{2\tan\left(\frac{W_{Eplane}}{2}\right)} \qquad (Eq\ 4)$$

The final and most critical dimension is the spacing of the metal plates. The blue Styrofoam sheets sold as insulation have excellent thickness uniformity, and $^3/_4$ inch is pretty near optimum for 10 GHz, but the actual dimension should be measured carefully. The thickness determines the index of refraction:

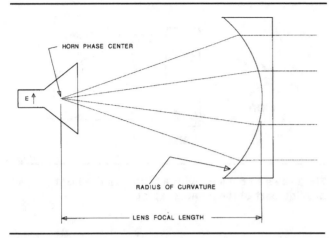

Fig 4—A single-curved lens. The radius of curvature is found using Eq 6, with the radius of the flat side set to infinity.

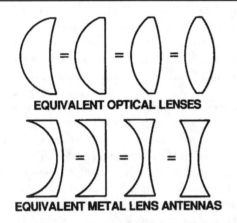

EQUIVALENT OPTICAL LENSES

EQUIVALENT METAL LENS ANTENNAS

Fig 5—Each of the lenses shown has the same focal length, per Eq 6.

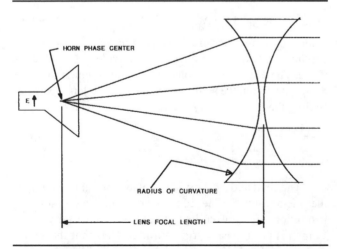

Fig 6—A double-curved lens. *HDL_ANT* provides both single-curved and double-curved lens designs.

$$\text{index} = \sqrt{1 - \left(\frac{\lambda_0}{2 \times \text{spacing}}\right)^2} \qquad \text{Eq 5}$$

which is the ratio of the wavelength in the lens to the wavelength in free space.

Next comes calculation of the lens curvature. The optimum curve is an ellipse, but we know that spherical lenses have been used for optics since Galileo, so a circle is a usable approximation. We can show that the circle is an excellent fit if the focal length is more than twice the lens diameter; photographers will recognize this as an $f/2$ lens. This suggests that the feed horn have a beam width of no more than 28°, or a horn aperture of at least two wavelengths.

The radius of curvature of the two lens surfaces is calculated from the lensmaker's formula (see Note 4):

$$\frac{1}{f} = (\text{index} - 1)\left(\frac{1}{R_1} - \frac{1}{R_2}\right) \qquad \text{Eq 6}$$

where a negative radius is a concave surface. For the single-curved surface of Fig 4, one radius is set to infinity. All combinations of R1 and R2 that satisfy the formula are equivalent, as shown in Fig 5; the computer program calculates the single-curved and symmetrical double-curved solutions (Fig 6). The radius of curvature calculated above is for the surface, and thus the central plate, which has the full curvature. The rest of the plates must be successively wider and have smaller radii so that the edges of all the plates form a spherical lens surface. This is more geometry, and the program does the calculations for each plate.

The final calculation involves the phase centers of the horn, so that the lens-to-horn distance matches the focal length. This is a difficult calculation involving calculation of Fresnel sines and cosines; KB1VC deserves credit for the programming.[9,10] Without a computer, you would use trial-and-error looking for best gain. What the calculations will show is that many horns, particularly the "optimum" designs, have much different phase centers in the E- and H-planes. The program offers to make a crude compensation for this, but, if possible, the H-plane aperture of the horn should be adjusted slightly to match the phase centers. A few trial runs of the program should enable you to find a good combination. If you already have a horn, either try the compensation or just use the E-plane phase center.

For very large lenses, the size may be reduced by stepping the width of the plates into zones which keep transmission in phase, as shown in Fig 7. The program will suggest a step dimension if it is useful. At 10 GHz, a step is useful only for lenses larger than 2 feet in diameter.

Construction

Construction is straightforward, using metal plates of aluminum foil spaced by Styrofoam, as suggested by HB9SLV (see Note 2). A 2-foot by 8-foot sheet of blue Styrofoam, 3/4-inch thick, is less than $5 at the local lumberyard and will make several antennas. The aluminum foil is attached to the foam using artist's spray adhesive, available at art supply stores. Spray both surfaces lightly, let them dry for a minute or two, then spread the foil smoothly on the

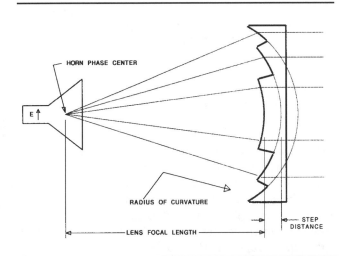

Fig 7—A zoned lens can be used to implement large lenses, reducing the needed thickness.

Fig 8—A 300-mm lens placed in front of a Gunnplexer transceiver provides about 10 dB of additional gain over that of the horn alone.

foam. If the adhesive melts the foam, you are using too much.

Next mark the outline of a rectangle for each metal plate on the foil. These will be used later to cut the foam and line up the plates, so they should all be the same size. Then mark the center of each curve and measure off the radius to the center of the circle. Using a compass with a hobby knife attached, place the point at the center of the circle and cut the curve through the foil into the foam. When all the curves are cut, peel off the unwanted foil, leaving the lens plates. Then cut up the rectangles with a razor blade and stack the blocks into a lens. (You did number them, didn't you?) Each rectangle should have foil on one side. If it looks good, glue them up two at a time. The final antenna will be a block of foam—there is no need to shape the foam to the lens curve. Shrink-wrapping the lens with thin plastic makes nice weatherproofing.

A few helpful hints are in order. Sharp knife blades really help in this process, and permanent markers don't smear. Also, if the foam is cut halfway through, it will snap cleanly on the line.

Adjustment

A metal-lens antenna only works in the E-plane. This is parallel to the elements of a dipole or Yagi but perpendicular to the wide dimension of a waveguide. The plates *must* be perpendicular to the wide dimension to provide gain.

The horn should point through the center of the lens, but the focus distance is not as critical as with a dish. Aiming is done by pointing the feed horn; the lens focuses the beam more tightly but does not change the beam direction. Tilting the lens will *not* steer the beam—if you don't believe this, take an optical lens, like a magnifying glass, focus it on something, and tilt it.

We found that the best gain was with the horn slightly closer to the lens than calculated, probably because of edge effects. Making the size of the plates slightly larger than calculated would probably eliminate this effect and make

the gain a bit higher; since a wavelength at 10 GHz is about an inch, an inch or two oversize is plenty.

One other interesting effect was found with Gunnplexers: since the transmitter is also the receiver local oscillator, reflected power from the lens adds to the LO power, or subtracts when out-of-phase. This makes the received signal strength vary with every half-inch change in lens-to-horn distance, with very little change in signal strength observed at the other station. So, adjust the spacing for the best received signal. Of course, this effect does not exist on a system with low LO radiation.

Using the *HDL_ANT* Program

The lens section of *HDL_ANT* calculates the dimensions for the plates of a lens. Since all curves are circles that are easily drawn with a compass, templates are not generated. The basic input data is entered interactively, then results are presented in tabular form. If you like the results, they may be saved to a file for printing or further processing; if not, try another run with new data.

All dimensions are in millimeters. There are two reasons for this: the first is that odd fractions lead to errors in measurement, and the second is that one millimeter is a good tolerance for 10-GHz lens dimensions. If all measurements are made to the nearest whole millimeter, good results can be expected. The only exception is in the plate spacing, and that is accurately controlled by the foam thickness.

Results

We have constructed and tested three metal-plate lens antennas to date: a 150-mm single-curved version, and 150-mm and 300-mm double-curved versions. Fig 8 shows the 300-mm lens fed by a WBFM Gunnplexer system, and Listing 1 shows the *HDL_ANT* design of this antenna. All the lenses are designed to be fed with the standard Gunnplexer horn, which has well-matched phase centers, whether by design or by accident. Gain measurements are shown in Table 1. The lenses perform with about 50% efficiency if we consider them as having a round aperture; the corners do not contribute significantly, but we made them square for convenient fabrication and mounting.

We also used the lenses during several contests during 1992, 1993 and 1994. The 300-mm lens increased the range of our WBFM Gunnplexer transceivers by approximately 50%, to over 200 km, enabling contacts over new paths. The equipment was still highly portable due to the light weight of the lens, and they have survived mishaps with only a few harmless dents in the foam.

Further Uses for Metal Lenses

The metal-lens antenna could be useful at other frequencies: for 5.76 GHz a foam thickness of around 35 mm would be good, and at 24 GHz approximately 8-mm-thick foam might work, though it could be lossy at that frequency.

A lens can also be part of a more complex antenna system. For instance, a divergent lens can be used to provide better illumination for some of the very deep dishes that are sometimes available as surplus. A book on optics (such as Note 4) will show how to change the focal points appropriately.

Lens Summary

We have demonstrated that metal-lens antennas may be easily designed and constructed using the *HDL_ANT* computer program and that a book-sized lens, light and rugged enough for backpacking, provides gain enhancement adequate to double the range of a Gunnplexer system.

Antenna Gain Measurement

Hams have been measuring antennas for many years at VHF and UHF frequencies, and we have seen marked improvement in antenna designs and performance as a result. Very few serious antenna measurements have been made at 10 GHz; the additional difficulties at these frequencies are not trivial. I'll describe a few new twists that make it more feasible.

Overview

Antenna measurement techniques have been well described by K2RIW and W2IMU;[11,12] the latter also appears in the *ARRL Antenna Book* and is required reading for anyone considering making antenna measurements.[13]

The antenna range is set up for antenna ratiometry so that two paths are constantly being compared, both originating from a common transmitting antenna. (See Note 11.) These two paths are called the "reference" and "measurement" paths. The reference path uses a fixed antenna that receives what should be a constant level. In reality, there are continuous small fluctuations in the received signal at microwave frequencies, even over a short distance like an antenna range. Using ratiometry, the reference path allows these random variations in the source power or the path loss to be corrected by an instrument that constantly compares the signal from the measurement path with that of this reference path. First, a standard antenna with known gain is

measured and the reading is recorded. Then, when an unknown antenna is measured, the difference between it and the standard antenna determines the gain of the unknown antenna.

Instrumentation

One measuring instrument commonly used is the venerable HP 416 ratiometer, with crystal detectors used to sense the RF. Basically, this technique compares the outputs of two crystal (diode) detectors. The crystal detectors present a problem: a matched pair is needed, and these are hard to find for 10 GHz. Also, diode detectors have poor sensitivity and dynamic range, so it is necessary to provide adequate power to keep the detectors operating in the square-law region where they are accurate. Another problem is drift in the old vacuum-tube HP 416.

It seemed to me that a superheterodyne technique was needed. If the signal could be converted to some lower frequency, it could be received on a better receiver. If the two channels had separate converters, the comparison could be made at the lower frequency. Finally, if we simply switched between the two channels at the lower frequency, the output would be an AM signal. If the switching rate were at an audio frequency, an AM receiver would thus have an audio output amplitude proportional to the difference in signal level between the two channels. Once the signals are combined by the switch, they may be easily amplified as needed at the lower frequency.

At 10-GHz, frequency stability is always a problem, so a normal communications receiver might be too sharp. From work with 10-GHz WBFM, I know that most signals are stable to within a few hundred kHz after warm-up, so a receiver with a 1-MHz bandwidth should be acceptable. While I was wondering if there might be something usable in a surplus catalog, it occurred to me that I already owned a perfectly usable solid-state wideband AM receiver—an AILTECH Model 75 Precision Automatic Noise Figure Meter (PANFI), which I found at a surplus auction. Not only that, it also has a synchronous detector and an output to synchronously switch the input signal (normally the noise source). The meter reads the difference in signal level as the input is switched; in this application, instead of the difference with a noise source switched on and off, it reads the difference as it switches between the two channels, with excellent resolution. If a signal much stronger than the noise is applied, the meter responds only to the signal rather than noise. (While checking the references for this paper, I discovered that K2RIW—see Note 11—had suggested use of the AILTECH Model 75 PANFI in 1976, but no one had remembered so I had to rediscover it!)

The only problem with the PANFI is that it is calibrated to solve the noise equation:

$$F_{dB} = T_{ex(dB)} - 10\log_{10}(Y-1) \qquad \text{(Eq 7)}$$

This requires a bit of arithmetic on a calculator or using the *HDL_ANT* computer program to undo the results and find the difference in dB:

Table 1—Summary of 10.368-GHz Antenna Measurements (N1BWT 12/18/93, 5/14/94, 9/15/94)

Antenna	Focal distance	Gain (dBi)	Efficiency
Standard Gain Horn, (22.5 dBi calculated)		22.45	43%
Scientific-Atlanta Model 12-8.2,			
courtesy KM3T, gain thanks to John Berry, Scientific-Atlanta.			
Gunnplexer Horn (17.45 dBi calculated)		17.5	57%
+ 6" lens	~8"	20.9	45%
+ 12" lens	~21"	27.4	50%
Surplus horn (19.4 dBi calculated)		19.6	67%
W1RIL loop Yagi		16.0	
22" dish, f/D = 0.39, surplus, feed = 11 GHz Superfeed:[*]			
unmodified feed	8.25"	33.1	55%
with feed line to reflector		32.2	45%
modified feed		32.9	53%
25" dish, f/D = 0.45, Satellite City, with the following feeds:			
11 GHz Superfeed[*]	10.875"	34.4	58%
Clavin feed	11.125"	34.1	54%
Rectangular Horn,			
E=0.9", H=1.38"	10.625"	33.7	49%
E=1.14", H=0.9"	10.625"	32.9	41%
WR-90 to coax Transition	11.0"	32.7	39%
WA1MBA log periodic	10.94"	32.4	37%
18" dish, f/D = 0.42, Satellite City, with the following feeds:			
11 GHz Superfeed[*]	7.75"	31.7	60%
Clavin feed	7.875"	31.2	53%
Rectangular Horn E=0.9", H=1.38"		31.5	57%
WR-90 to coax Transition, rect flange		30.2	42%
WR-90 to coax Transition		30.2	42%
round flange, od = 2.15"			
Cylindrical horn with	7.875"	~28[**]	~26%[**]
slotted choke ring to choke ring			
WA3RMX Triband feed		~17[**]	
24" Commercial (Prodelin) dish antenna:			
feed is rectangular horn fed by WR-90 waveguide "shepherd's crook"			
		33.6	52%

RANGE LENGTH = 102 feet. $2D^2/\lambda$ = 91 feet. Test height ~ 8 feet.

FOCAL DISTANCE SENSITIVITY: each feed was adjusted for max gain.
    Gain was down 1 dB about ¼" either way from peak.

Notes:[*]11 GHz Superfeed is a Chaparral feed horn for 11-GHz TVRO.
    [**]These feeds were not positioned accurately—more gain is possible.

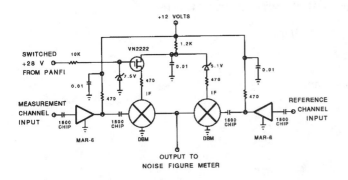

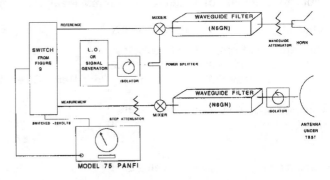

Fig 9—This switch can be used to automatically switch between the reference and measurement paths of the antenna gain measurement system.

Fig 10—Test setup for 10-GHz antenna measurements.

$$Y_{dB} = 10\log_{10}\left(1 + 10^{\left(T_{ex(dB)} - F_{dB}\right)/10}\right) \qquad \text{(Eq 8)}$$

Otherwise, the indicated gains are very optimistic.

The input to most noise figure meters is at 30 MHz, so I used a surplus signal generator to generate an LO 30 MHz away from 10368 MHz and used my 10-GHz transverter as the source transmitter. The signal generator provides the LO for two surplus waveguide diode mixers, but a pair of mixers like the ones in the transverter would also be fine.[14] Matched mixers aren't needed since they are linear mixers with wide dynamic range. I preceded each

mixer with a band-pass filter, but there probably aren't many stray 10-GHz signals around. An isolator in the antenna line is useful when the antennas may have high SWR.

Everything after the mixers is at 30 MHz, so ordinary cables and components complete the setup. I included a step attenuator in the measurement path to double-check the meter readings.

The switch, shown schematically in Fig 9, uses a common double-balanced mixer (DBM) as an attenuator in each path. Applying a dc current though the diodes in a DBM varies the attenuation; the DBM has high loss with no dc current, and low loss with dc current applied; I measured 54 dB of loss at 0 mA and 2.8 dB at 20 mA. An FET and some Zener diodes provide a crude switching circuit to switch the

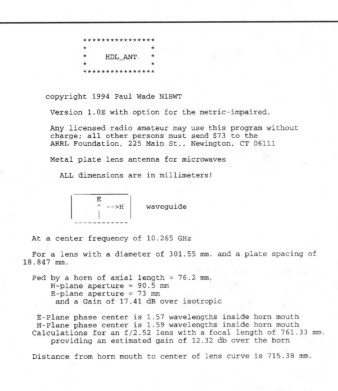

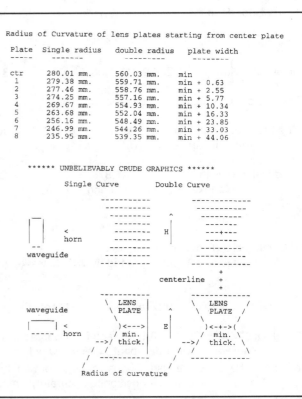

Listing 1

current in response to the 28-V output from the PANFI.

Fig 10 shows the antenna measuring setup for 10 GHz. The reference path uses a small horn as the receiving antenna, and the source antenna is another horn. After completing all connections, the signal generator is adjusted for maximum received signal, as indicated with the PANFI switched to the noise OFF position. Then the PANFI is switched to AUTO to display the difference between the paths, which is converted to relative gain using the above equation.

Antenna Range

The length of the antenna range is important: if it is too short, there will be significant phase difference over the aperture of the antenna being tested, resulting in low measured gain. The minimum range length to avoid this error is the Rayleigh distance:

$$\text{Rayleigh distance} = \frac{2D^2}{\lambda} \qquad \text{Eq 9}$$

A few trial calculations will show that this requires miles of range for large dishes. Fortunately, the Rayleigh distance for the 25-inch dish that I wanted to measure is only 91 feet at 10 GHz.

The antenna range is a ground-reflection range, as shown in Fig 11, where the range is designed to account for ground reflection and control it. One alternative would be to place the antennas high enough that ground reflection would be insignificant; however, in order to keep the reflected signal contribution from the ground to less than 0.5 dB, both ends of the range would have to be 122 feet high for a range length of 91 feet. Another type of range requires the signal path to be at a 45° angle to the ground, so the antenna height would only be 91 feet. For most amateur work, antenna heights like these are impractical, so the ground-reflection range is used.

In order to have the phase error as low in the vertical plane as in the horizontal plane, the height of the antenna being measured must be at least *four* times its aperture diameter, which is 100 inches for the 25-inch dish.[15] I suspect that most amateur antenna ranges have had insufficient antenna height and consequently have had trouble measuring higher-gain antennas accurately. My first measurements, at a height of about 4 feet, showed lower than expected efficiency for the larger dishes. Raising the height made the measured efficiency greatest for the larger dish, as you would expect, since the feed horn blocks a smaller percentage of the aperture.

The received energy should be at a maximum at the height of the antenna being measured. For a ground-reflection antenna range, this is controlled by the height of the source antenna:

$$h_{\text{source}} = \frac{\lambda}{4}\left(\frac{\text{Range length}}{h_{\text{receiving}}}\right) \qquad \text{Eq 10}$$

which works out to about 3 inches for the 91-foot-long range with the receiving antenna 100 inches high. Therefore, by adjusting the reference and measurement antennas to over 8 feet high (easily done on the back of a pickup truck or a

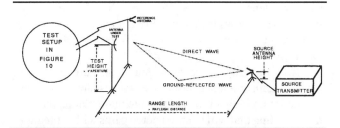

Fig 11—A ground-reflection antenna range with ratiometry. The required range length and antenna heights depend on the frequency and the characteristics of the antennas.

porch) and placing the height of the source and antenna around 3 inches, good, reliable and accurate results may be obtained (see Table 1).

The Standard-Gain Antenna

In order to measure meaningful antenna gains, an antenna with known gain is required. Recall that all measurements are relative to a known standard. A dipole is useless as a standard—its broad pattern receives so many stray reflections that repeatable readings are nearly impossible, and its gain is so much lower than a 30+ dB dish that equipment accuracy is a problem; few instruments are accurate over a 30-dB (1000:1 power ratio) range.

What is required is an antenna with a known gain, preferably a gain of the same order of magnitude as the antennas to be measured. At microwave frequencies, the gain of a horn antenna can be calculated quite accurately from the physical dimensions. The algorithm used in the *HDL_ANT* program will be accurate within about 0.2 dB if good construction techniques are used.

For even better accuracy, several companies make standard-gain horns with good calibration data. For 10 GHz, a standard-gain horn was lent to me by KM3T—he was lucky enough to find one surplus. Mr. John Berry of Scientific-Atlanta was kind enough to provide the gain calibration curve.

Measurements

Once the antenna range is designed and set up, it must be checked out before making actual measurements. This is best done with an antenna with a fairly broad pattern, like a medium-sized horn, as the test antenna. First, the attenuators are adjusted for a convenient meter reading. Then the field uniformity is probed by moving the test antenna horizontally and vertically around the intended measurement point. The indicated gain should peak at the center and should not vary significantly over an area larger than any antenna to be tested; the variation should be less than 1 dB. At this point, the height of the source antenna usually needs to be adjusted to get the vertical peak at the intended receiving height. Finally, the test antenna is held stationary, and calibrated attenuation steps are added in the test path to make sure the indicated gain (after correction if using a PANFI) changes by the amount of attenuation added. With a ratiometer, the attenuation must be added at the microwave frequency, but a PANFI system like Fig 10, with a linear mixer, allows the use of an IF attenuator; step attenuators are much easier to find (or build) for 30 MHz than for 10 GHz.

Now the range is ready to make measurements. The standard-gain antenna is inserted as the test antenna, aimed for maximum indication, and the attenuators adjusted for a meter reading that will keep expected gains within the range of the meter. All gain measurements will be the difference from this standard reading added to the gain of the standard-gain antenna. The standard-gain antenna is replaced by an antenna to be tested, the new antenna is aimed for maximum gain, and its indicated gain is recorded. The difference between this indicated gain and the standard reading, after correction, added to the known gain of the standard-gain antenna, is the gain of the test antenna. The reading with the standard-gain antenna should be checked frequently to correct for instrumentation drift; ratiometry with the reference antenna corrects for other sources of drift.

Measurement Results

I set up a 10-GHz antenna range in my yard that was 102 feet long, more than the Rayleigh distance, with the equipment described above. The received field was probed for uniformity, and the height of the source antenna was adjusted for a flat field at the required height. Then I was able to start measurements, using a standard-gain horn for comparison. The results are shown in Table 1.

The first thing that became apparent is that all adjustments on a dish are critical. In the field, looking at a tiny S-meter, it doesn't seem so difficult to point a dish with a beamwidth of only about 3 degrees. The PANFI, however, has a large meter with 1 dB expanded out to nearly an inch. On this meter, even tiny adjustments have obvious effects, demonstrating how touchy aiming a dish is.

The most critical dimension of a dish is the focal length—the axial distance from the feed to the center of the dish. A change of $^{1}/_{4}$ inch, or about a quarter-wavelength, changed the gain by a dB or more.

The critical focal length suggests that it is crucial to have the phase center of the feed exactly at the focus of the reflector. Since the phase center is rarely specified for a feed horn, we must determine it empirically by finding the maximum gain on a reflector with known focal length, which we can estimate from templates for various f/D. Thus we can estimate the phase centers for all the feeds in Table 1.

For the Chaparral style feed horns, we can deduce some further information. Several different feeds were measured, with two different dimensions, and with adjustable choke rings. Regardless of where the choke ring was set, maximum gain occurred with the choke ring the same distance from the reflector. This implies that the phase center is controlled by the position of the choke ring, not the central waveguide. The version designed for 11-GHz TVRO use, with the gain shown in Table 1, has an apparent phase center in front of the choke ring, while a larger one, dimensioned for 10 GHz, has the apparent phase center behind the choke ring (inside the ring) and provides gain similar to the smaller one.

As for efficiency, none of the dish measurements in Table 1 exceeds 60%, and it is obviously easy to get efficiencies less than 50%. This suggests to me that the 55% quoted in the books is far from typical, and careful design and measurement is needed to reach or exceed it. As illustrated in Part 2 of this series, dishes with small f/D (less than 0.3) may be very difficult to feed efficiently.

On the other hand, several of our amateur feeds have higher efficiencies than the commercial dish antenna shown in Table 1. If you find a surplus dish with a feed, don't assume it is the best possible one—different applications may require optimization of other parameters. For instance, WA1MBA has been working on a broadband log-periodic feed. The efficiency at 10 GHz shown in Table 1 is rather poor by comparison, but having a single dish feed that offers reasonable performance at several amateur microwave bands is an exciting possibility.

Several of the dish measurements in Table 1 were made with a coax-to-waveguide transition as a feed—the open-ended waveguide flange acts as a small horn. This is not an optimum feed, as shown by the low efficiency, but it is one that is readily available for comparisons. If the feed for your dish does not perform significantly better than a plain waveguide flange, it can certainly stand improvement.

Measured horn and lens efficiencies are comparable to dish efficiencies, so we can conclude that all three types of antennas can provide the same gain for the same aperture area. This leaves us free to choose the type of antenna best suited to the application.

Conclusion

Horns, dishes and lenses are all high-performance microwave antennas well-suited for amateur communications. Horns are small, rugged, and reliable, good for rover operation; they may be supplemented with a lens acting as an "amplifier" for increased gain. Dishes offer the ultimate in gain, at the expense of size and narrow beamwidth.

Horn and lens construction is easily within amateur capability, but parabolic reflectors at microwave frequencies require construction accuracy that is difficult to achieve. A dish antenna using a manufactured reflector still requires careful attention to detail to realize high efficiency.

Amateur antenna gain measurement at 10 GHz with good results has been demonstrated using ratiometry, and a noise figure meter is a good solid-state replacement for a vacuum-tube ratiometer. Antenna gain measurements are valuable for making critical adjustments and for verifying that an antenna is providing the performance expected. Better antenna gain measurements should bring the same improvement to amateur microwave antennas that years of antenna measuring contests have brought to VHF and UHF antennas.

Acknowledgments

I'd like to thank Bob Egan, N1BAQ; Larry Filby, K1LPS; Dick Knadle, K2RIW; Barry Malowanchuk, VE4MA; Dave Pascoe, KM3T; Matt Reilly, KB1VC; Ken Schofield, W1RIL; Dan Thompson, N1IOL; and Tom

Williams, WA1MBA, for their help, and Beth Wade, N1SAI, and Filomena Didiano for locating obscure references. I'd also like to remember the late Dick Turrin, W2IMU, who generously shared his vast knowledge of antennas and antenna measurement with many hams—he taught me more than I'll ever remember.

Notes

[1]Wade, P., N1BWT, and Reilly, M., KB1VC, "Metal Lens Antennas for 10 GHz," *Proceedings of the 18th Eastern VHF/UHF Conference*, ARRL, May 1992, pp 71-78.

[2]Vilaseca, Angel, HB9SLV, "Microwave Lens Antennas," *VHF Communications*, 3/90, pp 179-189.

[3]Kock, Winston E., "Metal-Lens Antennas," *Proc. I.R.E.*, Nov 1946, pp 828-836.

[4]Hecht, Eugene, *Optics*, Addison-Wesley, 1987.

[5]Franke, John M., WA4WDL, "Pour an Antenna for X-band," *73*, Feb 1991.

[6]Sletten, Carlyle J., W1YLV, *Reflector and Lens Antennas*, Artech House, 1988.

[7]We will not attempt to explain this seeming magic here. Refer to "phase velocity" in any good book on microwaves. Rest assured that no laws of physics are being violated.

[8]Kraus, John, W8JK, *Antennas*, McGraw Hill, 1956.

[9]Muehldorf, Eugen I., "The Phase Center of Horn Antennas," reprinted in Love, A. W., *Electromagnetic Horn Antennas*, IEEE Press, 1976.

[10]Abramowitz, Milton and Stegun, Irene A., *Handbook of Mathematical Functions*, Dover, 1972.

[11]Knadle, R. T., K2RIW, "UHF Antenna Ratiometry," *QST*, Feb 1976, pp 22-25.

[12]Turrin, D., W2IMU, "Antenna Performance Measurements," *QST*, Nov 1974, pp 35-41.

[13]Hall, G., K1TD, ed, *The ARRL Antenna Book*, ARRL, 1991, pp 27-43 to 27-48 (Also in the 1994 Edition.)

[14]Wade, P., N1BWT, "Building Blocks for a 10 GHz Transverter," *Proceedings of the 19th Eastern VHF/UHF Conference*, ARRL, May 1993, pp 75-85.

[15]Gillespie, E. S., "Measurement of Antenna Radiation Characteristics," in *Antenna Handbook: theory, applications, and design*, Y.T. Lo and S.W. Lee, editors, Van Nostrand Reinhold, 1988, page 32-22.

More on Parabolic Dish Antennas

Offset dishes, penny feeds and sun noise

By Paul Wade, N1BWT

(From *QEX*, December 1995)

Introduction

In the past year, you have probably noticed little gray dish antennas sprouting from rooftops and appearing for sale in stores as part of satellite TV systems. One common version is the RCA DSS system, which uses an 18-inch offset-fed dish.

In previous articles about parabolic dish antennas, I described only conventional axial-feed dishes because other types weren't readily available.[1,2] Now, with the introduction of the DSS system this is no longer true—inexpensive offset-feed dishes are readily available, and they offer excellent performance at 10 GHz.

This article is the "fourth part" of my three-part series of *QEX* articles on practical microwave antennas.[1,3,4] In order to show how to use offset dishes effectively, some familiarity with antenna terminology and concepts is required, so I urge the reader to review the earlier articles. In addition to offset-feed dishes, this article will also discuss the "penny" feed for conventional dishes, dishes with multiple reflectors and the use of sun noise to verify antenna and system performance.

[1] Notes appear at the end of this section.

Offset-Feed Dishes

An offset-feed dish antenna has a reflector that is a section of a normal parabolic reflector, as shown in Fig 1. If the section does not include the center of the dish, none of the radiated beam is blocked by the feed antenna and support structure. For small dishes, feed blockage in an axial-feed dish causes a significant loss in efficiency. Thus, we might expect an offset-feed dish to have higher efficiency than a conventional dish of the same aperture.

In addition to higher efficiency, an offset-feed dish has another advantage for satellite reception. The dish in Fig 2, aimed upward toward a satellite, has its feed horn pointing toward the sky. A conventional dish would have the feed horn above it, pointing toward the ground, as shown in Fig 3. Any spillover from the feed pattern of the conventional dish would receive noise from the warm earth, while spillover from the offset dish would receive less noise from the cool sky. Since a modern low-noise receiver, such as a satellite TV LNB, has a noise temperature much lower than the earth, the conventional dish will be noisier. This is *G/T*, which I described in the previous series of articles; the offset dish offers higher gain, *G*, since the efficiency is higher,

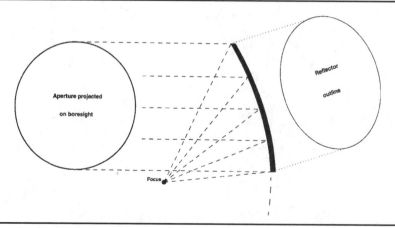

Fig 1—Geometry of an offset parabolic dish antenna.

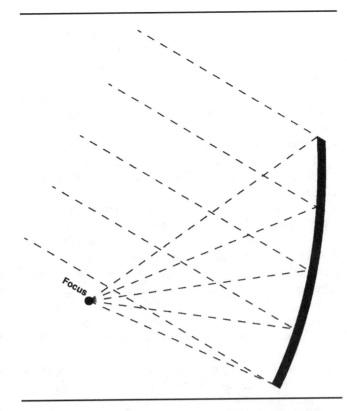

Fig 2—An offset parabolic dish antenna aimed at a satellite.

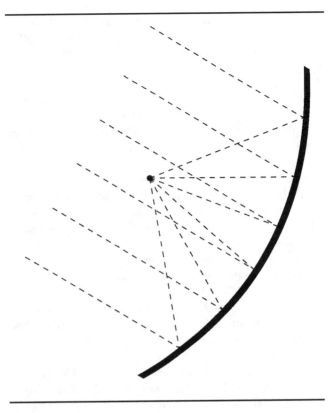

Fig 3—An axial-feed parabolic dish antenna aimed at a satellite.

plus reduced noise temperature, T, so both terms in the G/T ratio are improved. The higher gain means more signal may be received from a source, and the lower noise temperature means that less noise accompanies it, so a higher G/T offers a higher signal-to-noise ratio.

The RCA DSS Dish

The original incentive to use an offset-feed dish was provided by Zack Lau, KH6CP, who pointed out that the 18-inch RCA DSS dishes are available by mail order for about \$13.[5] I ordered a dish and a mounting bracket to see if I could figure out how to use one at 10 GHz.[6] When it arrived, it wasn't obvious where the feed point should be, so I took a trip to a local discount store to eyeball the system on display.

Now I had an idea where to put the feed, but not the exact location. The RCA reflector is oval shaped, but Ed, W2TTM, provided the needed insight: the dish aperture should appear circular when viewed on boresight, as shown in Fig 1. Thus the dish must be tilted forward for terrestrial operation. Although the reflector is an oval, the effective antenna aperture is the projected circle, with a diameter equal to the small dimension of the oval, 18 inches for the RCA dish. The tilt angle, feed point location and the rest of the dish geometry can be calculated—see the Appendix for the procedure. Version 2 of the *HDL_ANT* computer program will do these calculations. This program is available from the ARRL BBS (860-594-0306) or via the Internet at **http://www.arrl.org/qexfiles/hdl_ant2.zip** or **ftp://ftp.arrl.org/pub/qex/hdl_ant2.zip**.

The calculations show the focal length of the RCA dish to be 11.1 inches. If the dish were a full parabola rather than just an offset section, the diameter would be about 36 inches, for an *f/D* of 0.30, which would require a feed with a very broad pattern. However, a feed horn need only illuminate the smaller angle of the offset section, a subtended angle of about 77°. This subtended angle is the same as a conventional dish with an *f/D* of 0.7, so a feed horn designed for a 0.7 *f/D* conventional dish should be suitable. Rectangular feed horns have been shown to work well with offset reflectors and are readily designed to illuminate an *f/D* this large.[7] I used G3RPE's graph for rectangular feed horn design and the *HDL_ANT* computer program to design suitable rectangular horns.[8,9] I made two of different lengths from flashing copper. Subsequently, I added an approximation to G3RPE's curves to version 2 of *HDL_ANT* so the program can design feed horns for both offset and conventional dishes as well as generate templates for them.

Since the actual reflector geometry has an *f/D* of 0.30, the focal distance should be quite critical. As explained in Part 2 of my previous *QEX* series, this dimension is the most critical for dish antenna performance—even more critical for reflectors with smaller *f/D*—so the phase center of the feed should be positioned within a quarter-wavelength of the focal point. The RCA dish must be tilted forward to an angle of 66.9° from horizontal for terrestrial operation with the beam on the horizon. In this orientation, the focal point is just below the lower rim of the dish, so the feed horn is out of the beam. To locate the focus accurately, I calculated the distance to both the top and bottom of the rim, tied a knot in a piece of string and taped the string to the rim so the knot was at the focus when the string was pulled taut, as shown in Fig 4. Then I made a sliding plywood holder for the feed horn, taped it in place and adjusted it so that the knot in the string was at the phase center of the horn, about 6 mm inside the mouth of the horn, shown in Fig 5. (For visibility, the string in the photograph is much heavier than the kite string I used so a small knot could locate the focus more accurately.) Materials aren't critical when they aren't in the antenna beam!

Where should the feed horn be aimed? On a conventional dish it is obvious—at the center. However, an offset feed is much closer to one edge of the dish, so that edge will be illuminated with much more energy than the opposite edge. I read an article that did a lengthy analysis of the various aiming strategies and then suggested that small variations have little effect, so aiming at the center of the reflector is close enough.[10]

After all this analysis, it was time to see if the offset dish really works. We (W1RIL, WB1FKF, N1BAQ, and N1BWT) set up an antenna range and made the measurements shown in Table 1. The RCA dish with a simple rectangular feed horn measured 63% efficiency at 10 GHz, significantly higher than we've *ever* measured with an 18-inch conventional dish. Varying the focal distance showed that the calculations were correct and that this dimension is critical. Fig 6 is a template produced by the *HDL_ANT* program for the rectangular feed horn that gave the highest efficiency, and Fig 7 is a photograph of the feed horn I made with the template.

The higher efficiency of the offset-feed dish is mainly due to reduced blockage by the feed and supporting structure. Fig 8 is a photograph of a conventional dish while measuring sun noise, so that the shadow of the feed demonstrates the actual area blocked—neither light nor RF energy from the sun is reaching the reflector. Fig 9 is a photograph of the RCA offset dish peaked on the sun to measure sun noise; note that the shadow of the feed is only a tiny area at the bottom edge. Remember that these feed horns provide a tapered illumination, so the energy illuminating the center of the reflector is typically 10 dB stronger than at the edge. Thus, central blockage in a conventional dish is *ten times* worse than the same area blocked at the edge of an offset dish, and the photographs clearly illustrate how much more blocked area there is in a conventional axial-feed dish.

Other Offset Feeds

A rectangular feed horn is fine for linear polarization, but what if we want circular polarization? One popular feed that works well with circular polarization is the W2IMU dual-mode feed. The published amateur versions are all for *f/D* in the range 0.55 to 0.6, but Dick's original article also described another version for a different *f/D*.[11] It should be possible to make one for the 0.7 *f/D* needed for the RCA offset dish, but that would require some experimentation (or computer modeling, if you have software available) for optimum performance.

The truly adventuresome could try a trimode feed de-

signed specifically for offset-fed dishes.[12] The math is daunting, and construction appears difficult, but I have seen one TVRO feed that may use this design.

Other Offset Dishes

I was given an offset-feed, 24-inch plastic dish with a cosmetic defect (and no other information). Measurements showed the geometry to be similar to the RCA dish, so the same feed horns would work fine. I was not able to support the feed as well on this dish, so the feed location may not have been optimum, but it still measured 61% efficiency at 10 GHz.

Two other types of offset dishes seem to be fairly common, so some will probably wind up in amateur hands eventually. Many automobile dealerships and discount stores have larger offset dishes, four feet or more in diameter, with a reflector that appears circular. The other type is another brand of TVRO system, with an oddly shaped dish about 3 feet across; the ones I've seen are marked "Primestar." I had a chance to look one over at a county fair, next to the tractor dealer. The reflector appeared to be wider than it was high, requiring a fairly wide feed angle. The feed horn had a curved plastic surface that could possibly be a molded lens.

If I were to acquire one of these reflectors, I would place it flat on the ground with the reflecting surface facing upward and fill it with water, which provides a level surface from which to take measurements. The water should fill an oval area reaching the top and bottom edges of the rim, but not the sides. Measuring this oval as described in the Appendix, and measuring the depth and location where the water is deepest, should be enough to calculate the offset geometry. The feed horn beamwidth would have to be broader from side-to-side than from top-to-bottom, but a rectangular feed horn can be designed to provide an asymmetric pattern.

Mounting an Offset Dish

To aim an offset dish at the horizon with the feed below the dish, the reflector must be tilted forward—66.9° from horizontal for the RCA dish. One way to accomplish this would be to mount it on a wedge cut at the correct angle, so that the bottom of the wedge can be mounted on a level surface or tripod. An alternative technique is to rotate the dish so that the feed is to the side, level with the center of the dish. In this configuration, the elevation uncertainty is eliminated, but an aiming device must be provided for azimuth. An accurate azimuth readout is a good idea for any dish, since aiming a narrow beam by eye is fraught with error. A settable compass rose with one-degree gradations works well for rover operations.

The Penny Feed

The "penny" feed has been used for years with good results. It consists of a metal disc, originally an old (pre-decimalization) English penny, at the end of a waveguide with slots in the broad wall of the waveguide. I built one to see how well it really works, using dimensions by G4ALN from the *RSGB Microwave Handbook, Volume 3*.[13] The only English coin I had of the right diameter was ten new pence

Fig 4—Locating feed point for offset dish using calculated string lengths.

Fig 5—Knot in string accurately locates phase center of feed horn at focal point of offset dish.

Table 1
Summary of 10.368-GHz Antenna Measurements (Measurements by N1BWT, W1RIL, WB1FKF and N1BAQ, 7/6/95)

Antenna	Focal Dist	Gain (dBi)	Efficiency
Standard-Gain Horn (22.5 dBi calculated gain)[1]		22.45	43%
WB1FKF hombrew horn		22.05	
25-in dish, f/D = 0.45, from Satellite City, with the following feeds:			
11 GHz Superfeed[2]	11.187 in	34.3	56%
	11.0 in	34.0	52%
11 GHz Superfeed, modified	11.187 in	34.6	61%
with central waveguide flush with outer rings.[2]			
G4ALN "penny" feed	10.375	33.0	41.5%
18-in dish, f/D = 0.42, from Satellite City, with the following feeds:			
Clavin feed	7.875 in	31.2	53%
18-in offset dish, RCA DSS steel, with the following feeds:			
Rectangular Horn, E=31.2 mm, H=41.1 mm, Length=20 mm			
	11 in[3]	32.0	63.5%
	11.25 in[3]	31.0	50%
Rectangular Horn, E=31.2 mm, H=41.1 mm, Length=10 mm			
		31.5	57%
Rectangular Horn, surplus, E=30.1 mm H=45.2 mm, Length=42 mm			
	11 in[3]	31.8	61%
24-in (WB1FKF) with the following feeds:			
11 GHz Superfeed with Styrofoam housing[2]		34.4	62%
WA1MBA log-periodic		28.0	14%
24-in offset dish, plastic, with the following feed:			
Rectangular Horn, E=31.2 mm, H=41.1 mm, Length=20 mm.			
	14.75 in[3]	34.3	61%
30-in dish, f/D = 0.45, (lighting reflector), with the following feeds:			
11 GHz Superfeed, modified	13.5 in	36.4	64%
with central waveguide flush with outer rings[2]			

Measurement specifications:
Range: Length = 150 feet. $2D^2/\lambda$ = 135 feet. Test height ≈10 feet.
Focal distance: Each feed was adjusted for maximum gain. Axial dish focal distances measured to outermost point on feed.

Notes:
[1]Scientific-Atlanta model 12-8.2. Antenna courtesy KM3T, gain thanks to John Berry of Scientific-Atlanta.
[2]11 GHz Superfeed is a Chaparral feed horn for 11-GHz TVRO.
[3]Offset dishes measured from bottom edge of dish to center of horn aperture.

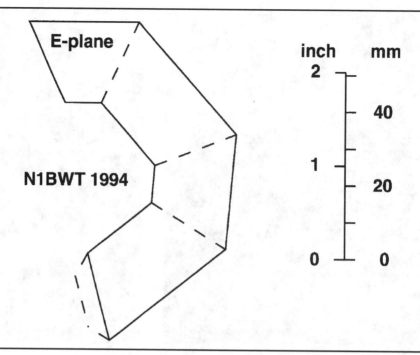

E-plane

N1BWT 1994

inch	mm
2	
	40
1	
	20
0	0

Fig 6—Template for a 10-GHz feed horn that can be used with the RCA DSS offset dish.

rather than an old penny, but silver should work at least as well as copper. The feed is easy to build, and has a good SWR, so I can see why it is popular. However, the performance was mediocre, with 41% efficiency, about the same as an open waveguide flange. Thus, the gain of a 25-inch dish fed with a penny feed is not much higher than the 18-inch offset dish fed with a simple horn.

To be fair, the dish we used, with an *f/D* of 0.45, is not optimum for the penny feed. The *Handbook* states that it is suitable for dishes with an *f/D* ratio in the range 0.25 to 0.3. A dish that deep is extremely difficult to illuminate well, so it is unlikely that this feed will deliver much higher efficiency than we measured. However, it is probably as good a feed as any for very deep dishes.

Cassegrain and Gregorian Feeds

Large professional antennas often use multiple reflector feeds, like the Cassegrain (hyperbolic subreflector) and Gregorian (elliptical subreflector) configurations.[14] Even better is a shaped-reflector system, where both reflector shapes are calculated for best efficiency and neither reflector is parabolic.[15] JPL reports 74.5% efficiency on their 34 meter high-efficiency antenna.[16]

All of these systems require a carefully shaped subreflector that is more difficult than a parabola to fabricate. For a shaped reflector to work well, it must be larger than 10 wavelengths, and the main reflector must be much larger than the subreflector to minimize blockage by the subreflector. One analysis suggested that a Cassegrain antenna must have a minimum diameter of 50 wavelengths, with a minimum subreflector diameter of 20 wavelengths, before the efficiency is higher than an equivalent dish with a primary feed.[17] This is a fairly large dish, even at 10 GHz, and shaping a 20 λ subreflector is beyond the ingenuity of most hams. However, there is probably a surplus one somewhere, and the scrounging ability of hams should never be underestimated.

Fig 7—Photograph of 10 GHz rectangular feed horn for RCA DSS offset dish made using template in Figure 6.

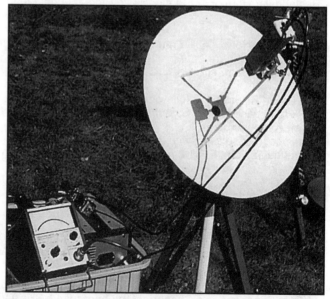

Fig 8—Conventional dish receiving sun noise. Shadow of feed horn demonstrates aperture blockage by feed.

Sun Noise Measurement

Even a modest 10-GHz system is capable of detecting sun noise, which is an excellent way of ensuring both antenna and receiver performance since we can predict how much sun noise should be received with a given antenna size and receiver noise figure.[18] Only a relatively simple setup is required to make reasonably accurate sun noise measurements.

On the other hand, setting up an antenna range to evaluate antenna performance, as described in my earlier articles, requires a significant amount of equipment and a good standard antenna of known gain, and it is still one of the most difficult measurements to perform accurately.

A good system for measuring sun noise was described by Charlie, G3WDG.[19] He built a 144-MHz amplifier with moderate bandwidth using MMICs and helical filters that amplifies a transverter output to drive a surplus RF power meter. The newer solid-state power meters, like the HP 432 and more recent models, are stable enough to detect and display small changes in noise level, and the response is slow enough to smooth out flicker. Since my 10-GHz system has an IF output at 432 MHz, duplicating Charlie's amplifier would not work. In the junk box I found some surplus broad-

Fig 9—Offset dish receiving sun noise. Feed horn shadow at edge of dish demonstrates minimal aperture blockage by offset feed.

band amplifiers and a couple of interdigital filters. I combined these to provide high gain with a few MHz-bandwidth, arranged as shown in Fig 10. The first filter limits the bandwidth so we are measuring at the desired frequency, 10.368 GHz in this case, and the second filter at the output is important to limit the noise bandwidth at the detector, since noise power is proportional to bandwidth. Without the second filter, the broadband noise generated by the amplifiers or MMICs would overwhelm the sun noise, whose bandwidth is limited by the first filter. Approximately 100 dB of total gain is required with a bandwidth of 10 MHz for an output power of one milliwatt. I found that roughly 60 dB of gain after my preamp and transverter was required to get a reasonable level on the power meter.

Operation is simple—point the dish at the sun, peak the noise, then move to clear sky and note the difference in output. Several precautions are necessary:

1. According to G3WDG, amplifiers with broadband noise output suffer gain compression at levels about 10 dB lower than found with signals, so be sure the amplifier compression point is at least 10 dB higher than the indicated noise.

2. Make sure no stray signals appear within the filter pass-band.

3. A clear area of sky is necessary, since foliage and other obstructions add thermal noise that can obscure the cold-sky reading. I found a large tree generated more noise than the sun because it filled the whole beam and appeared in sidelobes as well. The measurement is really comparing sun noise plus all other noise to all the other noise received, so stray sources of thermal noise can produce error. Fortunately, this error is almost always in the pessimistic direction, so we aren't led astray.

4. If the preamp is at or near the feed, don't let it heat up too much or its noise temperature can change. (Total solar radiation is about one kilowatt per square meter—that's several hundred watts on even a small dish.)

Before making measurements, I used the *NOISE* program by Mel, WRØI, to estimate expected sun noise.[18] For a 2-foot dish with 60% efficiency and a receiver noise figure of about 2.5 dB (modified TVRO LNB), the program predicted 2.4 dB of sun noise. My initial measurements using the setup described below showed 2.5 dB of sun noise on my 25-inch dish and 2.0 dB with the 18-inch, offset-fed RCA dish. However, I also measured 2.2 dB of sun noise on a 30-inch dish with a fancy "shepherd's crook" feed arrangement using copper water pipe as circular waveguide. The last mea-

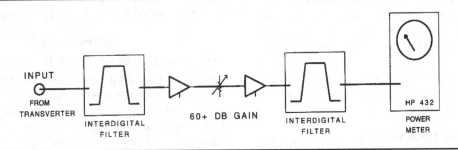

Fig 10—Block diagram of an indicator for sun-noise measurements.

Table 2: More Sun Noise Measurements

22 Oct 1995, 1:00 PM

Antenna	Sun noise
Standard Gain Horn	0.35 dB
30-inch dish, mod. Chaparral feed	3.2 dB
25-inch dish, Chaparral feed	2.3 dB
18-inch dish, Clavin feed	1.6 dB
18-inch offset—steel, rect. horn feed	2.4 dB
18-inch offset—SMC, rect. horn feed	2.5 dB

Note: Estimated noise figure = 2.9 dB

surement quickly highlighted the need for further adjustment of the feed arrangement. The version of the *NOISE* program I used only calculates for dish sizes in integral feet, so we can't get precise estimates for these small dishes, but an improved version is available.

After I made the equipment more portable and stable, I was able to measure sun noise on most of my 10-GHz antennas, with the results shown in Table 2. The *G/T* advantage of the offset dish for satellite communication is clearly demonstrated: the 18-inch offset dish is not only much better than the equivalent size conventional dish, but outperforms a 25-inch conventional dish that has 2 dB more gain, as shown in Table 1.

Sun-noise measurements are fine for checking system performance but less satisfactory for making adjustments. Any adjustment may change both sun and sky reading, so it is necessary to compare the two after each adjustment, and the resulting differences may be small. Make one adjustment at a time, keep careful notes and look for reproducible improvements. The process is tedious, but careful work pays off.

If you've never tried it, you are probably wondering why you can't just use your receiver to measure sun noise. The answer is that you can, but with less accuracy and more frustration because of the narrow bandwidth and short time constant of a communications receiver. First, the noise-measuring equipment described above has a bandwidth of a few MHz, while a typical receiver bandwidth is 3 kHz, a thousand times narrower. To compensate for a thousand times narrower bandwidth, a thousand times more gain, or 30 dB more, is required. Most receivers have adequate gain but use AGC to control the gain; if you can't turn off the AGC, a problem with many receivers, the audio output doesn't change linearly with input level, and the S-meter is far too small to resolve tenths of a dB. With the AGC off, the audio output follows the input noise, but the narrow bandwidth and short time constant (about one millisecond, limited by the lowest audio frequency response, typically 300 Hz) produce an output with fluctuations caused by the random nature of noise—I've typically seen one dB of flicker, making it hard to read tenths of a dB. With the power meter, the thermistor sensor has a time constant of hundreds of milliseconds, which smoothes and averages the flicker to produce a very stable meter indication.

Receiver Noise Figure Using the Sun Noise Equipment

The same equipment used for measuring sun noise can also be used to measure receiver noise figure. While measuring sun noise, I noticed that pointing an antenna at the ground produced a significant noise increase. I then realized that this is similar to a hot/cold system for noise figure measurement, where the earth is about 290 K while the cold sky at 10 GHz is around 6 K at high elevations, so the temperature difference is nearly 290 K.[20] Using the standard-gain horn, I found approximately 3 dB of difference between cold sky and warm earth; I had previously measured this LNB preamp at 2.9 dB of NF, or just under 290 K of noise temperature, so a 3-dB increase is exactly right, as shown by the following calculations.

The difference between the hot and cold noise sources is called the *Y* factor; this is used to calculate the receiver noise temperature, T_e, as follows:

$$T_e = \frac{T_{ground} - Y \cdot T_{sky}}{Y - 1}$$

where Y is a power ratio (convert from dB).

The noise temperature is easily converted to noise figure, F, if you prefer:

$$F(dB) = 10 \cdot \log\left(\frac{T_e}{290} + 1\right)$$

This technique should work with any antenna with reasonably high gain and low sidelobes, so stray noise is minimized. A long horn is a good choice. Just point the antenna at clear sky overhead, away from the sun or any obstruction, note the meter reading, then point the antenna into the ground and read the noise increase Y. For convenience, I've added these calculations to version 2 of *HDL_ANT*.

Azimuth Alignment Using the Sun

Computer programs are available that will calculate the sun's azimuth and elevation at a given place and time, so peaking on the sun can be used to calibrate both azimuth and elevation readout. For a rover without a computer, a previously calculated list giving azimuth at half-hour increments at expected rover locations is useful for setup in each location. Don, WB1FKF, suggests that if you are unable to measure sun noise, a vertical line on the dish will suffice on sunny days; simply line up the feed horn's shadow on the line.

Recommendations for Parabolic Dish Feeds

Table 3 is an update of the recommendations I made for dish feeds in previous articles. The numbers shown are my best estimates for small dishes at 10 GHz, and the recommendations should be taken as my personal opinion only. See the previous *QEX* articles for the appropriate references.

Conclusion

The new DSS offset-feed dishes are readily available small microwave dishes, and I have shown how to use them as high-performance 10-GHz antennas. Their high performance, convenient size and low cost should make them the antenna of choice for portable operation.

Sun-noise measurement capability is a valuable tool for measuring and verifying performance of both antennas and

Table 3—N1BWT Recommendations for Dish Feeds

Type of Feed	f/D Optimum	Best η estimate	η for f/D=0.45	Comments
Chaparral	0.35-0.45	55-65%	61-64%	"11 GHz Superfeed" from Chaparral dealers good at 10 GHz
VE4MA/Kumar	0.35-0.45	55-65%	61%	proven performance at 1296, 2304, and 3456 MHz
W2IMU Dual-Mode	0.5-0.6	55-60%	NR	proven performance 432 MHz to 10 GHz
Rectangular horn	>0.45	50-60%	58%	tailor dimensions for f/D —also good for Offset dishes
Clavin	0.35-0.4	50-60%	57%	small feed blockage
EIA Dual-dipole	0.5-0.6	50-60%	NR	better at lower frequencies
Circular horn	Function of diameter	25-50%	26%	asymmetrical E- and H-planes and phase centers
Penny (G4ALN)	0.25-0.3	30-45%	41.5%	attractive mechanically
Dipole	0.3-0.4	30-45%	NR	asymmetrical E- and H-planes
Log Periodic	?	10-40%	14%	broadband, but poor phase centers

Measurements and Calculations for an Offset Parabolic Reflector

The geometry of an offset-feed dish antenna is a bit more complicated than a conventional dish antenna, but the measurements needed to use one are straightforward. We need to first determine the tilt angle of the reflector, then do some curve fitting calculations for the dish surface, calculate the focal length and finally determine the focal point in relation to the offset reflector.

One common type of offset parabolic reflector has an oval shape, with a long axis from top to bottom and a shorter axis from side to side. However, if you were in the beam of this antenna, looking down the boresight, it would appear to be circular, with the feed at the bottom. Tilt the top of the reflector forward, until it appears circular from a distance, and it will be in the correct orientation to operate with the beam on the horizon. The tilt can be determined much more accurately with a simple calculation:

Tilt angle (from horizontal) = arcsin (short axis/long axis) [Note: the arcsin function is called sin^{-1} on some scientific calculators.]

For the RCA 18-inch dish, the short axis is 460 mm (about 18 inches) and the long axis is 500 mm. Therefore, the tilt angle = arcsin (460/500) = 66.9° above horizontal. At 10 GHz, one millimeter is sufficiently accurate for most dish dimensions, so using millimeters for calculations eliminates a lot of tedious decimals.

If the offset reflector is not oval, we can still use the same calculation by placing it on the ground with the reflecting surface upward and filling it with water; the surface of the water is a level plane from which to make measurements. The surface of the water in the dish should be an oval just touching the top and bottom rims, while the other axis of the oval of water is the shorter axis.

The other dimension we need is location and depth of the deepest point in the dish. The deepest point is probably not at the center, but somewhere along the long axis. Using a straightedge across the rim for an oval dish, or the water depth for other shapes, locate the deepest point and measure its depth and distance from the bottom edge on the long axis.

For the RCA dish, the deepest point is 43 mm deep at 228 mm from the bottom edge on a line across the long axis.

When the dish is tilted forward to 66.9° above horizontal, the translated coordinates describe the curve of the long axis by three points:

0, 0 mm	(bottom edge)
49.8, 226.6 mm	(deepest point)
196, 460 mm	(top edge)

If we assume that the bottom edge is not at the axial center of a full parabola of rotation (the equivalent conventional dish of which the offset dish is a section), but rather is offset from the center by an amount X_0, Y_0, then all three points must fit the equation:

$$4 * f * (X + X_0) = (Y + Y_0)^2$$

The unknowns are X_0, and Y_0, and f, the focal length; plugging in the three points gives us three equations and three unknowns, a readily soluble 3x3 matrix (actually, the 0,0 point allows reduction to a 2x2 matrix, even easier, followed by a simple calculation for X_0 and Y_0). Version 2 of the **HDL_ANT** program will do the calculations for you.

For the RCA dish, the answers are:

f = 282.8 mm = 11.13 inches
X_0 = 0.1 mm behind bottom edge
Y_0 = 11 mm below bottom edge, so the feed doesn't block the aperture at all.

So, we tilt the dish to 66.9° from horizontal, and the feed is on a line 11 mm below the bottom edge of the dish. To help locate the focal point, it is 283 mm from the bottom edge and 479 mm from the top edge, both edges on the long axis. I tied a knot in a piece of string and taped it to the top and bottom edges so that the knot locates the focal point.

For the RCA dish, we can also calculate the illumination angle to be 77° on the long axis and 79° on the short axis, so it is roughly symmetrical. The optimum feed for this illumination angle is equivalent to an axial-feed dish with $f/D \cong 0.7$.

Although the illumination angle is equivalent to an $f/D \cong 0.7$, the surface is a section of a parabola about 37 inches in diameter with a focal length of about 11 inches. Thus, the real f/D is 0.3, so the focal distance is quite critical.

receivers, and for antenna alignment. Also, since it is much easier to achieve accurate results with sun noise than with traditional antenna-range measurements, the various VHF conferences might consider using sun noise for antenna measurement.

Notes

[1] Section 2 of this book, pp 10-19.

[2] Wade, P. C., N1BWT, "High-Performance Antennas for 5760 MHz," *QEX,* January 1995, pp 18-21.

[3] Section 1 of this book, pp 1-9.

[4] Section 3 of this book, pp 20-28.

[5] Lau, Z., KH6CP/1, "RF," *QEX,* March 1995, p 24.

[6] MCM Electronics, 650 Congress Park Drive, Centerville, OH 45459, 800-543-4330. Steel dish part number 221196 or 221197, mounting bracket part number 221199. SMC dish part number 221198, mounting bracket part number 221200.

[7] Huang, J., Rahmat-Samii, Y. and Woo, K., "A GTD Study of Pyramidal Horns for Offset Reflector Antenna Applications," *IEEE Transactions on Antennas and Propagation,* AP-31, March 1983, pp 305-309.

[8] Evans, D., G3RPE, "Pyramidal horn feeds for paraboloidal dishes," *Radio Communications,* March 1975. (Also in Note 9.)

[9] Dixon, M. W., G3PFR, *Microwave Handbook, Volume 3,* RSGB, 1992, p 18.85.

[10] Jamnejad-Dailami, V. and Rahmat-Samii, Y., "Some Important Geometrical Features of Conic-Section-Generated Offset Reflector Antennas," *IEEE Transactions on Antennas and Propagation,* AP-28, November 1980, pp 952-957.

[11] Turrin, R. H., W2IMU, "Dual Mode Small-Aperture Antennas," *IEEE Transactions on Antennas and Propagation,* AP-15, March 1967, pp 307-308.

[12] Rudge, A. W. and Adatia, N.A., "Offset-Parabolic-Reflector Antennas, A Review," *Proceedings of the IEEE,* December 1978, pp 1592-1618.

[13] Dixon, M. W., G3PFR, *Microwave Handbook, Volume 3,* RSGB, 1992, pp 18.85-18.86.

[14] Hannan, P. W., "Microwave Antennas Derived from the Cassegrain Telescope," *IRE Transactions on Antennas and Propagation,* AP-9, March 1961, pp 140-153.

[15] Williams, W. F., "High Efficiency Antenna Reflector," *Microwave Journal,* July 1965, pp 79-82.

[16] Rafferty, W., Slobin, S. D., et al, "Ground Antennas in NASA's Deep Space Telecommunications," *Proceedings of the IEEE,* May 1994, pp 636-645.

[17] Rusch, W. V. T., "Scattering from a Hyperboloidal Reflector in a Cassegrainian Feed System," *IEEE Transactions on Antennas and Propagation,* AP-11, July 1963, pp 414-421.

[18] Graves, M. B., WRØI, "Estimating Sun Noise at Various Frequencies, Based on the 10.5 cm Flux Reported by WWV," *Proceedings of Microwave Update '94,* ARRL, 1994, pp 125-131.

[19] Suckling, C., G3WDG, "144 MHz wideband noise amplifier," *DUBUS,* 2/1995, pp 5-8.

[20] Graves, M. B., WRØI, "Computerized Radio Star Calibration Program," *Proceedings of the 27th Conference of the Central States VHF Society,* ARRL, 1993, pp 19-25.

The W3KH Quadrifiliar Helix Antenna

If your existing VHF omnidirectional antenna coverage is just okay, this twisted 'tenna is probably just what you need!

By Eugene F. Ruperto, W3KH

(From *QST*, August 1996)

I still remember that hollow, ghostly signal emanating from my receiver in 1957. The signal was noisy and it faded, but that was to be expected—it was coming from outer space. I couldn't help but marvel that mankind had placed this signal sender in space! They called it *Sputnik*, and it served to usher in the space race.

Little did I realize then that four decades later we would have satellites in orbit around Earth and other heavenly bodies performing all sorts of tasks. Now we tend to take satellites for granted. According to the latest information on the Amateur Radio birds, I count about 15 low-Earth-orbit (LEO) satellites for digital, experimental and communications work, and two in *Molniya*-type highly elliptical orbits (AO-10 and AO-13), with the probability of a third to be launched in early 1997.

The world has access to several VHF weather satellites in low Earth orbit. Unlike geostationary Earth-orbiting satellites (GOES), the ever-changing position of the LEOs presents a problem for the Earth station equipped with a fixed receiving antenna: signal fading caused by the orientation of the propagated wavefront. This antenna provides a solution to the problem. Although this antenna is designed primarily for use with the weather sats, it can also be used with any of the polar-orbiting satellites.

These days, technical advances and miniature solid-state devices make it relatively easy for an experimenter to acquire a weather-satellite receiver and a computer interface at an affordable price. So it was only a matter of time before I replaced my outdated weather-sat station with state-of-the-art equipment.

Yesterday

In the early '70s, I built a drum recorder that used a box with a light-tight lid. It was a clumsy affair. The box and photo equipment took up most of the 6×8-foot room in which it was housed. Next to the recorder, a 3×4-foot table supported a tube-type receiver, frequency converters, a reel-to-reel tape recorder (our data-storage medium), a 50-pound monitor oscilloscope, az/el rotator controls for the helical

antennas and a multitude of other devices including the drum-driver amplifiers and homemade demodulator. This station provided coverage of the polar-orbiting and geostationary satellites and furnished me with "tons" of data. Over time, my weather-satellite station evolved into a replica of mission control for the manned-spaceflight program! I had so much gear, it had to be housed in a shed separate from the house.

Today

Now, my entire weather-satellite station sits unobtrusively in one corner of the shack, occupying an area of less than one square foot—about the same size as my outboard DSP filter. My PC—now the display for weather-sat photos—is used for many applications, so an A/B switch allows me to toggle the PC between the printer and the weather-satellite interface.

What I needed next was a simple antenna system for unattended operation—something without rotators—something that would provide fairly good coverage, from about 20° above the horizon on an overhead pass. It was a simple

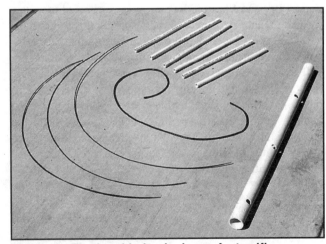

Figure 1—The humble beginnings of a terrific antenna.

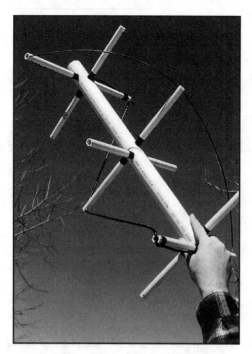

Figure 2—The quadrifilar helix antenna with two of the four legs (filars) of one loop attached.

Figure 3—This view shows the QHA with all four legs in place. The ends of the PVC cross arms that hold the coaxial leg are notched; the wire elements pass through holes drilled in the ends of their supporting cross arms.

request, but apparently one without a simple solution.

Background

Initially I used a VHF discone antenna with mixed results. The discone had a good low-elevation capture angle, but exhibited severe pattern nulls a few minutes after acquisition of signal and again when the satellite was nearly overhead. The fades and nulls repeated later as it approached the other horizon. About this time, Dave Bodnar, N3ENM (who got me reinterested in the antenna project), built a turnstile-reflector (T-R) array. The antenna worked fairly well but exhibited signal dropout caused by several nulls in the pattern. Dave built two more T-Rs, relocating them for comparison purposes. Unfortunately, the antennas retained their characteristic fades and nulls. Another experimenter and I built T-Rs and we experienced the same results. I suggested that we move on to the Lindenblad antenna. The Lindenblad proved to be a much better antenna for our needs than either the T-R or the discone, but still exhibited nulls and fades. Over a period of several months, I evaluated the antennas and found that by switching from one antenna to another on the downside of a fade, I could obtain a fade-free picture, but lost some data during the switching interval. Such an arrangement isn't conducive to unattended operation, so my quest for a fade-free antenna continued.

The Quadrifilar Helix Antenna

Several magazines have published articles on the construction of the quadrifilar helix antenna (QHA) originally

[1] Notes appear at the end of this section.

developed by Dr. Kilgus,[1] but the articles themselves were generally reader unfriendly—some more than others. One exception is *Reflections* by Walt Maxwell, W2DU.[2] Walt had considerable experience evaluating and testing this antenna while employed as an engineer for RCA.

Part of the problem of replicating the antenna lies in its geometry. The QHA is difficult to describe and photograph. Some of the artist's renditions left me with more questions than answers, and some connections between elements as shown conflicted with previously published data. However, those who have successfully constructed the antenna say it is *the* single-antenna answer to satellite reception for the low-Earth-orbiting satellites. I agree.

Design Considerations

I had misgivings about the QHA construction because the experts implied that sophisticated equipment is necessary to adjust and test the antenna. I don't disagree with that assumption, but I *do know* that it's possible to construct a successfully performing QHA by following a cookbook approach using scaled figures from a successful QHA. These data—used as the design basis for our antennas—were published in an article describing the design of a pair of circularly polarized S-band communication-satellite antennas for the Air Force[3] and designed to be spacecraft mounted. Using this antenna as a model, we've constructed more than six QHAs, mostly for the weather-satellite frequencies and some for the polar-orbiting 2-meter and 70-centimeter satellites with excellent results—*without the need for adjustments and tuning*. Precision construction is not my forte, but by following some prescribed universal calculations, a reproducible and satisfactory antenna can be

Figure 4—Another view of the QHA.

Figure 5—An end-on view of the top of the QHA prior to soldering the loops and installing the PVC cap.

built using simple tools. The proof is in the results.

The ultrahigh frequencies require a high degree of constructional precision because of the antenna's small size. For instance, the antenna used for the Air Force at 2.2 GHz has a diameter of 0.92 inch and a length of 1.39 inches! Nested inside this helix is a smaller helix, 0.837 inch in diameter and 1.27 inches in length. In my opinion, construction of an antenna *that* size requires the skill of a watchmaker! On the other hand, a QHA for 137.5 MHz is 22.4 inches long and almost 15 inches in diameter. The smaller, nested helix measures 20.5 by 13.5 inches; for 2 meters, the antenna is not much smaller. Antennas of this size are not difficult to duplicate even for those of us who are "constructionally challenged" (using pre-cut pieces, I can build a QHA in *less than an hour!*).

Electrical Characteristics

A half-turn half-wavelength QHA has a theoretical gain of 5 dBi and a 3-dB beamwidth of about 115°, with a characteristic impedance of 40 Ω. The antenna consists basically of a four-element, half-turn helical antenna, with each pair of elements described as a *bifilar,* both of which are fed in phase quadrature. Several feed methods can be employed, all of which appeared to be too complicated for us with the exception of the infinite-balun design, which uses a length of coax as one of the four elements. To produce the necessary 90° phase difference between the bifilar elements, either of two methods can be used. One is to use the same size bifilars, which essentially consist of two twisted loops with their vertical axes centered and aligned,

and the loops rotated so that they're 90° to each other (like an egg-beater), and using a quadrature hybrid feed. Such an antenna requires *two* feed lines, one for each of the filar pairs.

The second and more practical method, in my estimation, is the self-phasing system, which uses *different-size loops:* a larger loop designed to resonate *below* the design frequency (providing an inductive reactance component) and a smaller loop to resonate *higher* than the design frequency (introducing a capacitive-reactance component), causing the current to lead in the smaller loop and lag in the larger loop. The element lengths are 0.560 λ for the larger loop, and 0.508 λ for the smaller loop. According to the range tests performed by W2DU, to achieve *optimum* circular polarization, the wire used in the construction of the bifilar elements should be 0.0088 λ in diameter. Walt indicates that in the quadrifilar mode, the fields from the individual bifilar helices combine in optimum phase to obtain unidirectional end-fire gain. The currents in the two bifilars must be in quadrature phase. This 90° relationship is obtained by making their respective terminal impedances $R + jX$ and $R - jX$ where $X = R$, so that the currents in the respective helices are −45° and +45°.

The critical parameter in this relationship is the terminal reactance, X, where the distributed inductance of the helical element is the primary determining factor. This assures the ±45° current relationship necessary to obtain true circular polarization in the combined fields and to obtain maximum forward radiation and minimum back lobe. Failure to achieve the optimum element diameter of 0.0088 λ results in a form of elliptical, rather than true circular polarization, and the performance may be *a few tenths of a decibel* below optimum, according to Walt's calculations. For my antenna, using #10 wire translates roughly to an element diameter of 0.0012 λ at 137.5 MHz—not ideal, but good enough.

To get a grasp of the QHA's topography, visualize the antenna as consisting of two concentric cylinders over

Table 1

Quadrifilar Helix Antenna Dimensions

Freq (MHz)	Wavelength λ (inches)	Leg Size (0.508 λ)	Small Loop Diameter (0.156 λ)	Small Loop Length (0.238 λ)	Leg Size (0.560 λ)	Big Loop Diameter (0.173 λ)	Big Loop Length (0.26 λ)
137.5	85.9	43.64	13.4	20.44	48.10	14.86	22.33
146	80.9	41.09	12.6	19.25	45.30	14.0	21.03
436	27.09	13.76	4.22	6.44	15.17	4.68	7.04

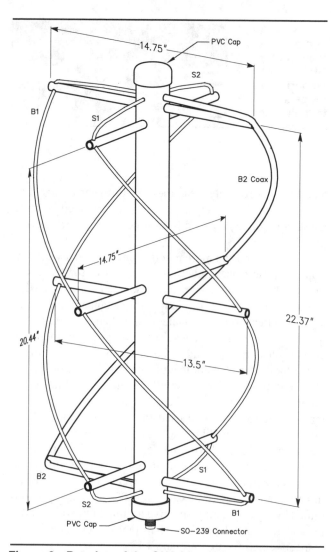

Figure 6—Drawing of the QHA identifying the individual legs; see text for an explanation. You may want to add an inch or two of PVC pipe at the bottom (and extend the coax to match) to make mounting easier.

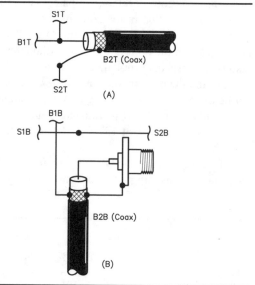

Figure 7—At A, element connections at the top of the antenna. B shows the connections at the bottom of the antenna. The identifiers are those shown in Figure 6 and explained in the text.

Figure 8—It's said that "The proof of the pudding is in the eating." To a weather-satellite tracker, clear, no-fade, no-noise pictures such as this one—compliments of W3KH's quadrifilar helix antenna—are delicious fare!

which the helices are wound (see Figures 1 through 5). In two-dimensional space, the cylinders can be represented by two nested rectangles depicting the height and width of the cylinders. The width of the larger cylinder (or rectangle) can be represented by 0.173λ, and the width of the smaller cylinder represented by 0.156λ. The length of the larger cylinder or rectangle can be represented by 0.260λ, and the length of the smaller rectangle or cylinder can be represented by 0.238λ. Using these figures, you should be able to scale the QHA to virtually any frequency. Table 1 shows some representative antenna sizes for various frequencies, along with the universal parameters needed to arrive at these figures.

Physical Construction

After several false starts using plywood circles and plastic-bucket forms to hold the helices, I opted for a simple PVC solution that not only is the simplest from a constructional standpoint, but also the best for wind loading. I use a 25-inch-long piece of schedule 40, 2-inch-diameter PVC pipe for the vertical member. The cross arms that support the helices are six pieces of 1/2-inch-diameter PVC tubing: three the width of the large rectangle or cylinder, and three the width of the smaller cylinder. Two cross arms are needed for the top and bottom of each cylinder. The cross arms are oriented perpendicularly to the vertical member and parallel to each other. A third cross arm is placed midway between the two at a 90° angle. This process is repeated for the smaller cylindrical dimensions using the three smaller cross arms with the top and bottom pieces oriented 90° to the large pieces. Using ⁵/₈-inch-diameter holes in the 2-inch pipe ensures a reasonably snug fit for the ¹/₂-inch-diameter cross pieces. Each cross arm is drilled (or notched) at its ends to accept the lengths of wire and coax used for the elements. Then the cross arms are centered and cemented in place with PVC cement. For the weather-satellite antennas, I use #10 copperclad antenna wire for three of the helices and a length of RG-8 for the balun, which is also the fourth helix. (I do not consider the velocity factor of the coax leg for length calculation.) For the UHF antennas, I use #10 soft-drawn copper wire and RG-58 coax. Copperclad wire is difficult to work with, but holds its shape well. Smaller antennas can be built without the cross arms because the wire is sufficiently self-supporting.

To minimize confusion regarding the connections and to indicate the individual legs of the helices, I label each loop or cylinder as B (for big) and S (for small); T and B indicate top and bottom. See Figures 6 and 7. I split each loop using leg designators as B1T and B1B, B2T and B2B, S1T and S1B and S2T and S2B, with B2 being the length of coax and the other three legs as wires. For right-hand circular polarization (RHCP) I wind the helices *counterclockwise* as viewed from the top. This is contrary to conventional axial-mode helix construction. (For LHCP, the turns rotate *clockwise* as viewed from the top.) See Figure 7 for the proper connections for the top view. When the antenna is completed, the view shows that there are two connections made to the center conductor of the coax (B2) top. These are B1T and S1T, for a total of three wires on one connection. S2T connects to B2T braid. The bottom of the antenna has S1B and S2B soldered together to complete the smaller loop. B1B and the braid of B2B are soldered together. I attach an SO-239 connector to the bottom by soldering the center conductor of B2B to the center of the connector and the braid of B2B to the connector's shell. The bottom now has two connections to the braid: one to leg B1B, the other to the shell of the connector. There's only one connection to the center conductor of B2B that goes to the SO-239 center pin.

Insulator Quality

A question arose concerning the dielectric quality of the tubing and pipe used for the insulating material. Antennas—being reciprocal devices—exhibit losses on a percentage basis, the percentage ratio being the same for transmit and receive. Although signal loss may not be as apparent on receive with a 2-μV signal as with a transmitted signal of 100 W (ie, it would be apparent if dielectric losses caused the PVC cross arms to melt!), signal loss could be a signifi-

cant factor depending on the quality of the insulating material used in construction. As a test, I popped the pipe into the microwave and "nuked" it for one minute. The white PVC pipe and the tan CPVC tubing showed no significant heating, so I concluded that they're okay for use as insulating materials at 137.5 MHz or thereabouts.

The antennas cost me nothing because the scrap pieces of PVC pipe, tubing and connectors were on hand. Total price for all new materials—including the price of a suitable connector—should be in the neighborhood of $8 or less.

Results

I use a 70-foot section of RG-9 between the receiver and antenna, which is mounted about 12 feet above ground. As with the earlier antennas, I use a preamp in the shack. With AOS (acquisition of signal) on the first scheduled pass of NOAA-14, I was pleasantly surprised to receive the first of many fade-free passes from the weather satellites, including some spectacular pictures from the Russian Meteors! Although the design indicates a 3-dB beamwidth of 140°, an overhead pass provides useful data down to 10° above the horizon. (My location has a poor horizon, being located in a valley with hills in all directions but south.) I've also received almost-full-frame pictures of the West Coast and northern Mexico at a maximum elevation angle of only 12° at my location. (The 70-cm antenna works fine for PACSATs, although Doppler effect makes manual tracking difficult.) The weather-satellite antenna prototype worked better than expected and a number of copies built by others required no significant changes. The quadrifilar helix antenna is *definitely* a winner! And believe me, *it's easy to build!*

Acknowledgments

Thanks to Chris Van Lint, and Tom Loebl, WA1VTA, for supplying me with the necessary technical data to complete this project. A special thanks to Walt Maxwell, W2DU, for his review and technical evaluation and for sharing his technical expertise with the amateur satellite community.

Notes

[1]C. C. Kilgus, "Resonant Quadrafilar Helix," *IEEE Transactions on Antennas and Propagation,* Vol AP-17, May 1969, pp 349 to 351.

[2]M. Walter Maxwell, W2DU, "Reflections, Transmission Lines and Antennas," (Newington: ARRL, 1990). [This book is now out of print.—*Ed.*]

[3]Randolph W. Brickner Jr and Herbert H. Rickert, "An S-Band Resonant Quadrifilar Antenna for Satellite Communication," RCA Corp, Astro-Electronics Division, Princeton, NJ 08540.

Photos by the author

Application of Circular Waveguide With an 11-GHz TVRO Feed

The circular waveguide (³/₄-inch copper type M) shepherd's crook feed described by WA6EXV in the San Bernardino Microwave Society's December 1993 newsletter was utilized in conjunction with a "Chaparral" brand 11-GHz TVRO Super- feed described by N1BWT. This feed system, with a 30-inch diameter, 0.375 F/D ratio, aluminum dish has been successfully used and has resulted in 2.8 dB of sun noise. This combination is being explored by the Long Island based TEN-X Group.

By Bruce Wood, N2LIV

(From *The 22nd Eastern VHF/UHF Conference*)

Crook

A sketch of the shepherd's crook feed is provided in Figure 1 with a listing of the pipe lengths utilized to construct it for this dish size, F/D ratio, and 11.25-inch focal length. These section lengths may be adjusted for various other size dishes. "NIBCO" brand pipe fittings were used for the elbows and couplings. The pipe lengths indicated includes the length of pipe recessed within the fittings.

Launcher

Several styles of SMA to round waveguide launchers were constructed as shown in Figure 1. The basic dimensions followed WA6EXV's design. Thread-in SMA connectors were used, Amphenol #901-9027. To gain more thread depth ¹/₂of a coupler sleeve was soldered on, or a small brass block constructed. The simplest method of launcher construction provided a rear wall for the waveguide and sufficient additional thread depth utilizing a ³/₄ inch "NIBCO" pipe end cap. The NIBCO pipe end cap technique is unpopular in some areas because of so called slightly unpredictable results. When soldering the end cap, make sure the pipe and cap is super cleaned, coated with liquid rosin flux, and be sure the solder "wicks" all the way to the bottom of the end cap. Failure to do this could result in a "microwave choke joint" that could make tune up more difficult.

Feed

The "Chaparral" brand Model #11-0148 feed horn was connected directly to the circular shepherd's crook feed by cutting off the existing waveguide flange on the Chaparral feed horn and enlarging with a lathe the existing remaining ³/₄" hole section to ⁷/₈". This will allow the ³/₄" copper pipe to be mounted directly within the feed to a depth of approximately ¹/₂". Anti oxidant grease was applied to help prevent corrosion between the copper and aluminum and the feed horn was finally epoxied in place.

Dish Mounting

The shepherd's crook was secured to the aluminum dish's center mounting plate with ordinary plumbing fittings. I originally planned to use a simple ³/₄-inch pipe coupling thru the dish's center plate but was concerned about the difficulty of soldering copper pipe fittings to the aluminum plate, possible galvanic corrosion in the salt air here on Long Island, and structural strength. I then located sweat to threaded screw type fittings which were also much stronger than a simple pipe coupling fitting and required no soldering. NIBCO brand fittings were used to construct the center dish feed-thru that will allow adjustment of the focal length and polarity. A pair of ³/₄-inch copper male and female adapters part #C604 & #C603, respectively, were reamed out to ⁷/₈-inch ID to allow the shepherd's crook to pass through them.

Two adjustable reamers from MSC Model 02239069 and 02239077 with a large tap handle were used to cut the hole. Approximately one hour was required to perform this operation. Screw the male threaded adapter pipe into the female before placing both into the vise, and performing the reaming operation. This supports the male adapter properly. The adapter becomes quite thin after the reaming. Be sure to insert a piece of ³/₄-inch pipe before applying a wrench to the male adapter. The ³/₄-inch pipe will keep it from deforming during the tightening operation. If a ³/₄-inch to 1-inch NIBCO threaded pipe adapter is used, the amount of reaming required is drastically reduced to approximately 10 minutes). In addition, the resulting couplings are much stronger. The slight disadvantage is that a large hole is required in the center of the dish. If a lathe is available this is a second option. The rear male adapter was slotted in four places and a stainless steel hose clamp was used to apply sufficient compressive forces as to not deform the shepherd's crook and to also secure the feed in place after adjustment of the focal length and polarity. Large washers may be used to take up any slop.

Results

The dish has a theoretical gain of 33.6 dBd and a 2.43 degree beam width. While on the antenna test range the polarity, focal length and coax to waveguide adapter (for phase and polarization rotation within the crook) were adjusted for maximum signal strength. Measurements on the antenna range were curtailed due to rain. Subsequent sun noise measurements resulted in 2.8 dB of sun noise, when using a 2.2 dB NF sun noise measurement instrumentation.

A Return Loss of better than 20 dB was obtained by launcher probe adjustment.

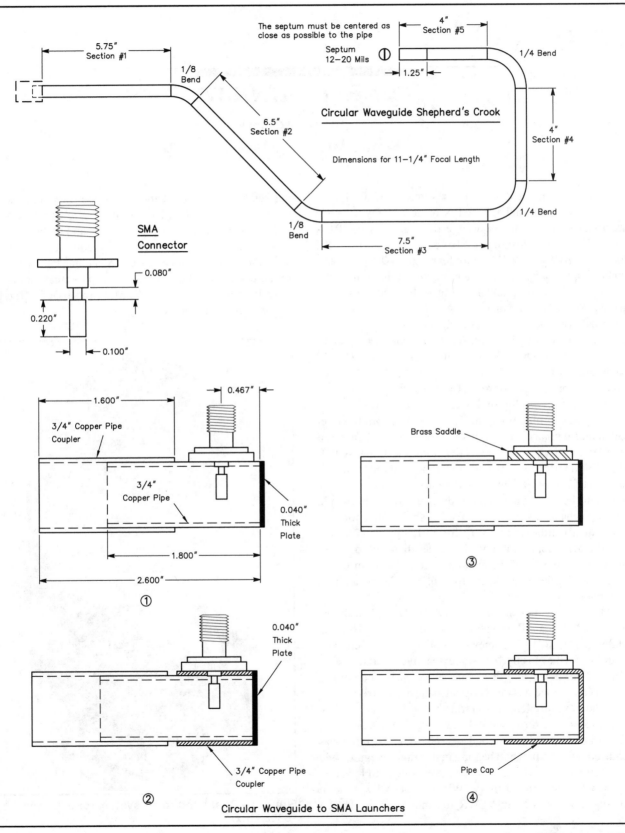

Circular Waveguide Shepherd's Crook

Dimensions for 11-1/4" Focal Length

SMA Connector

Circular Waveguide to SMA Launchers

Dual-Band Feedhorn for the DSS Offset Dish

5760 and 10368 MHz

By Paul Wade, N1BWT

n1bwt@qsl.net

(From *Microwave Update '97*)

I recently completed a new transverter for 5760 MHz in a fairly small package. It fits on top of my 10 GHz transverter next to the wedge that supports the RCA DSS offset dish. I designed a 5760 MHz feed horn for the dish using my *HDLANT21* computer program (**http://www.arrl.org/qexfiles**), built one, and modified the transverter slightly to allow for quick changing of the feed horns with two wingnuts. Now I had a package, shown in Figure 1, for a compact two-band rover station.

I was wondering if it was possible to make a good dual-band feed when Dick, K2RIW, mentioned that WR-112 waveguide covers both 5760 MHz and 10368 MHz; even though the handbooks don't list it as usable for 5760, the cutoff frequency is slightly lower so it still works.

The next problem was designing a feed horn to cover both bands with decent illumination for the dish. A few trial calculations showed that a 10 GHz horn providing –10 dB edge illumination taper would provide a –3 dB edge illumination at 5760 MHz—most of the energy would miss the dish! On the other hand, a horn designed for 5760 would have a much narrower beam at 10 GHz, so the outer portions of the dish would receive very little illumination energy; only the performance of a much smaller dish would result. After some fiddling of the numbers, I found a compromise which might have the same loss of efficiency at both frequencies.

The final design, using the *HDLANT21* template shown in Figure 2, has an illumination taper of roughly –16 dB at 10.368 GHz, so it is somewhat under-illuminated, and roughly –5 dB at 5760 MHz, somewhat over-illuminated. I adjusted the horn length to match the phase centers at 10.368 GHz, since it is most critical at the higher frequency.

The next problem was getting a good VSWR at both frequencies. The surplus WR-112 waveguide-to-coax transitions I had weren't very good at 5760 MHz, so tuning was required. I put a small ball bearing inside the waveguide and moved it around with a magnet on the outside until I located a spot which improved the VSWR at 5760 MHz without making the 10368 MHz VSWR too much worse. Then I

marked the spot, drilled and tapped the waveguide, and put in a tuning screw. Next I adjusted the screw for best VSWR at 5760 MHz, then put the BB back in and looked for a spot that improved both frequencies. A second screw was added here, then both screws adjusted for a compromise with reasonable VSWR at both frequencies. The final tuning had a VSWR under 1.6 at both 5760 MHz and 10368 MHz, but it is *not* a broadband match.

Figure 1—Dual-band rover system for 5760 and 10368 MHz.

Does it work? YES!

I completed it just in time for sun noise measurements at the July 1997 N.E.W.S. meeting, and tested it there on 10368 MHz. The DSS dish with a single-band horn feed has an efficiency better than 60%, while the dual-band feed is around 50%; the gain difference works out to about 1.2 dB.

The next day, I set up a sun noise measurement at 5760 MHz, with similar results: the DSS dish with a single-band horn feed has an efficiency of about 60%, while the dual-band feed is around 50%; the gain difference works out to about 1 dB on this band.

Summary

An RCA DSS dish with this dual-band feed horn provides two band performance only 1 dB down from a single band feed horn on each band. I've never seen a multiband feed with performance this good. This compact antenna is ideal for rover operations.

Questions

Q — Is a tri-band feed horn possible?

A — Not with ordinary waveguide, which cover a frequency range of less than 2 to 1 between cutoff and an upper frequency where other modes can propagate. Ridged waveguide can cover a wider range, but the horn design involves even more compromises.

Q — Is a dual-band horn possible for lower bands?

A — Yes, with a larger offset dish. A dish should be at least 10 λ in diameter for good performance, so the 18 inch RCA dish isn't big enough below 5760 MHz.

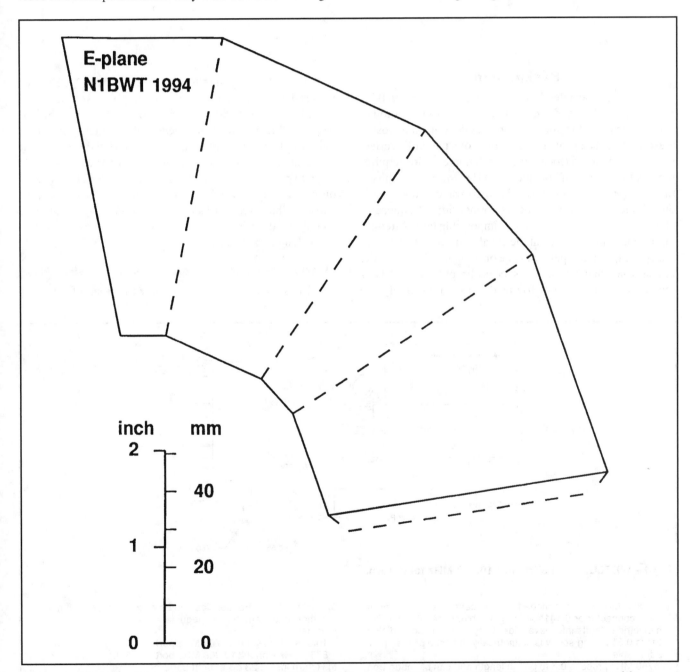

**E-plane
N1BWT 1994**

inch   mm
2
        40
1
        20
0   0

Figure 2—Dual-band feed horn template for RCA DSS offset dish; 5760 and 10368 MHz, WR-112 waveguide. Phase center is about 5 mm inside horn.

Dual-band Feed Horns for 2304/3456 MHz and 5760/10,368 MHz

By Al Ward, WB5LUA

(From *1997 Central States VHF Conference Proceedings*)

Background

Numerous articles have been written by WA9HUV, VE4MA, N1BWT and others on the proper illumination of a parabolic reflector. Joel Harrison has documented most of these works.[1] The proper illumination of a parabolic reflector with a given F/d (focal length to diameter ratio) requires the careful balance of both the E and H plane beamwidths of the feed horn. The problem on the microwave frequencies is one of putting several feed horns for individual frequencies at the same focal point—a nearly impossible task. Attempting to put multiple feeds at the focal point of the dish generally compromises performance on all bands. The satellite industry has had reasonable success by putting a 12 GHz feed in the middle of a 4 GHz feed. This is most likely due to the significantly smaller diameter of the 12 GHz feed versus the 4 GHz feed. With the relatively closer spacing of the 2304, 3456, 5760, 10,368 MHz bands this technique becomes difficult. Multiple feeds that are slightly offset are one way of obtaining multiband operation but there are some disadvantages, such as pointing offsets for each band. In order to get around the offset pointing problem I began work on in-line feeds, which will be the subject of this article. Any multiband feed will have compromises but I believe the techniques described herein will still result in a high performance antenna system.

Early Experiments on 2304 and 3456 MHz

I first experimented with inline multiband feeds back

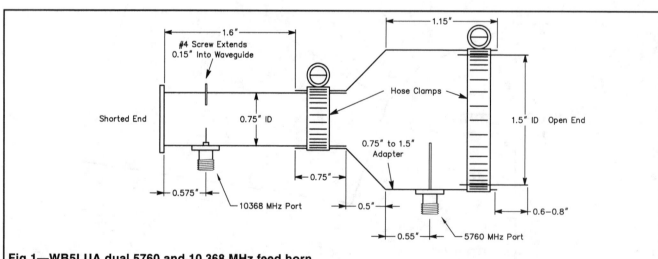

Fig 1—WB5LUA dual 5760 and 10,368 MHz feed horn.

Notes
1) 10,368 MHz probe is made from the center conductor of an SMA connector or 0.141" semi-rigid cable. 0.07" of the Teflon dielectric extends into waveguide. Length of pin above dielectric is 0.3" Tuning screw is diametrically opposite probe and is adjustable.
2) 5760 MHz probe is 0.6 to 0.7" in length and can be made from tubing 0.07 to 0.1" in diameter.
3) Tuning of both frequencies can be accomplished by tuning either probe length or waveguide length.
4) Isolation:
   10,368 MHz signal@5760 MHz port = −19 dB
   5760 MHz signal@10,368 MHz port = −45 dB
5) Return loss < 23 dB at both ports

in 1989 when I wanted a 2304 and 3456 feed that could be placed at the focal point of the dish and not require an offset in pointing between bands. I got the idea for the inline feed after analyzing the single band dual mode W2IMU feed, which has been used successfully on 1296, 2304 and 10,368 MHz, primarily for EME. The W2IMU feed has two different diameter circular waveguide sections which are designed to equalize the resultant E and H-plane beamwidths. The equal E and H-plane beamwidths with the appropriate taper contribute to a well illuminated high gain antenna. My thought was, what about feeding the larger outer section on the next lower amateur band? I decided to apply this concept to a dual band feed horn for 2304 and 3456 MHz. I used a standard 4 inch coffee can for 2304 MHz followed by a standard soup can for 3456 MHz. The results were very encouraging. This feed has been duplicated by several people over the years including K2DH, AA5C and W5ZN with good results. The construction of this feed and performance on a 32 inch dish is covered in detail in Joel Harrison's article.

Adding 5760 MHZ to Make a Three-Band Feed

I wanted to add 5760 to the original 2304/3456 MHz feed so I decided what would be easier than to just add a 1.5 inch diameter copper pipe to the end of the 3456 MHz can. The results were mixed. Yes, the horn worked but as I found out, the gain was considerably lower than theoretical. This was probably due to the fact that with the large aperture of the multiband feed at 5760 MHz, the feed was under-illuminating the dish.

Separate Dual Band Feed

I decided that the optimum combination would be to just duplicate the 2304/3456 feed for 5760 and 10,368 MHz. The result actually looks very similar to a W2IMU feed for 10,368 MHz. The resultant feed horn, shown in Figure 1, worked very well on 5760 MHz and was only slightly lower than expected on 10,368 MHz. The feed was tried on several dishes with varying F/d ratios and diameters. The resultant antennas were tested during a recent North Texas Microwave Society antenna workshop hosted by Kent Britain, WA5VJB. The results are documented in Table 1.

Test Results

Starting at 5760 MHz, the dual band feed worked very well, producing gains within a dB or two of theoretical 55% numbers when installed on 48 and 55 inch solid dishes and 55 and 72 inch perforated dishes. The new dual band 5760/10,368 MHz feed actually had 6 dB greater gain on 5760 MHz than did the original three band feed as measured on the same 55 inch dish.

On 10,368 MHz, the numbers were down a little but the 72 inch perforated dish, which was the only dish rated for 12 GHz, was still measuring 40.7 dBi. I did not optimize the actual position of the feed. The feeds were placed with the focal point slightly in the mouth of the feed.

The dual 2304/3456 MHz feeds were tested in the

same dishes but were slightly offset as only the dual 5760/10,368 MHz feed was at the focal point. As the results show, the gain numbers were somewhat lower than expected but the antenna range was only about 125 ft long and it could be that the larger dishes were underilluminated for the tests.

Construction

The length of both circular waveguide sections was made variable in order to improve the tunability of the feed horn. The monopoles can be preset as shown in Figure 1 and final tuning if needed can be accomplished by tuning the length of the waveguides. The resultant isolation between bands is very good and allows each band to be individually tuned. See Figures 2 and 3. The very good isolation also

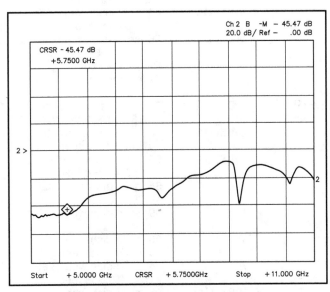

Fig 2—WB5LUA dual 5760 and 10,368 MHz feed horn —5760 MHz port to 10,368 MHz port isolation.

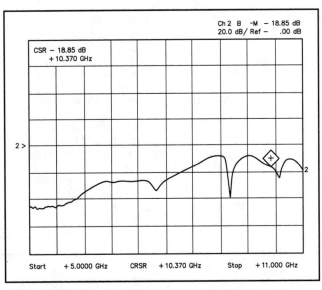

Fig 3—WB5LUA dual 5760 and 10,368 MHz feed horn— 10,368 MHz port to 5760 MHz port isolation.

Table 1
1997 NTMS Antenna Gain Measuring Party

Conducted by WA5VJB on March 23, 1997

Compiled by WB5LUA

Antenna range may have been too short for larger dishes, as gain numbers appear compressed.

Band (MHz)	Call	Design	Gain (dBi)	Theoretical 55% Gain (dBi)
1296	KA5BOU	15 el Yagi	16	
2304	WB5LUA	72" perf dish with coffee can feed	27	30
	AA5C	6' 40 el Yagi	20.4	
	WB5LUA	55" solid dish with coffee can feed	24.4	27
	WA5VJB	Reference horn	13.5	
3456	WB5LUA	72" solid dish with dual 2304/3456 feed	27.9	34
	WB5LUA	48" solid dish with offset soup can feed	25.7	30
	WB5LUA	55" solid dish with dual 2304/3456 feed	23.9	31
	WB5LUA	DEM loop Yagi	19.5	
	WA5VJB	Reference horn	16.9	
5760	WB5LUA	72" perf dish with dual 5760/10,368 feed	37.0	38
	WB5LUA	48" solid dish with 1.5" diam copper feed	33.5	34
	WB5LUA	55" perf dish with dual 5760/10,368 feed	33.0	35
	WB5LUA	55" solid dish with dual 5760/10,368 feed	32.5	35
	W5ZN	39" solid dish with scalar feed	31.3	32
	WA5TKU	30" solid dish with 1.5" diam copper feed	27.5	30
	AA5C	24" solid dish with dual 5760/10,368 feed	27.5	28
	WB5LUA	55" solid dish with old WB5LUA 3-can feed	27.0	35
	WB5LUA	12"×18" horn	21.0	
	WA5VJB	Reference horn	15.5	
10,368	WB5LUA	72" perf dish with dual 5760/10,368 feed	40.7	43
	WB5LUA	55" solid dish with dual 5760/10,368 feed	38.7	41
	WB5LUA	55" perf dish with dual 5760/10,368 feed	37.7	41
	WA5TKU	30" solid dish with 1.5" diam feed	33.7	36
	AA5C	24" solid dish with dual 5760/10,368 feed	33.2	34
	W5ZN	24" solid dish with WR90 to scalar feed	32.5	34
	WB5LUA	18" fiberglass dish with WR90 feed	28.7	31
	WA5VJB	Reference horn	17.7	

minimizes the additional isolation required in order to keep from destroying the front-end of the receiver for the other band. I believe part of the increased success of the 5760/10,368 MHz feed horn in regards to low frequency to high frequency isolation may, in part, be due to the smoother transition from the small section to the large section. Secondly it could be due to the 5760 MHz port having a poorer return loss at 10,368 MHz. Be aware that there are several different types of 0.75" to 1.5" transitions available and all may tune slightly differently.

Conclusion

I am very encouraged by the initial results of the multi-band feeds. I now have one dish for 2304, 3456, 5760 and 10,368 MHz. The 5760 and 10,368 MHz feed is at the focal point with the 2304/3456 MHz feed slightly offset. End result is that if 5760 MHz is peaked on a particular station then 10,368 MHz is also peaked. Same is true of 2304 and 3456 MHz. Good luck. Feedback is greatly appreciated.

Note

[1]"Horns for the Holidays", *1997 Proceedings of the Central States VHF Society Conference*, p 53-63.

TRANSVERTERS

BUILDING A 6-CM TRANSVERTER
By Zack Lau, W1VT

3456-MHz TRANSVERTER
By Zack Lau, W1VT

2304 AND 3456 NO-TUNE TRANSVERTER UPDATES
By Steve Kostro, N2CEI

MODERNIZING THE 3456 MHz NO-TUNE TRANSVERTER
By Jim Davey, WA8NLC

HOME-BREWING A 10-GHz SSB/CW TRANSVERTER (PARTS 1 AND 2)
By Zack Lau, W1VT

AN IMAGE-PHASING TRANSVERTER FOR 10.368 GHz
By Doug McGarrett, WA2SAY

Building a 6-cm Transverter

By Zack Lau, W1VT

(From *QEX*, June 1996)

Here are some ideas and circuits you can use to build your high-performance 6-cm transverter. I'll describe my latest mixer and band-pass filter designs and provide references for building the entire transverter. I'll also discuss my approach to integrating everything into a useable system.

5760-MHz Mixer

The mixer for 6 cm is quite similar to the one I designed for 10 GHz, using a hybrid splitter and a pair of $^6/_4$-wavelength ring mixers etched on Rogers 15-mil 5880 Duroid. 6 cm is perhaps the ideal band for using these mixers. The size is just about right—on 3 cm the rings are a little too tiny and on 9 cm they are getting a bit too big. More important, use of a relatively low 144-MHz IF makes it easy to design the hybrid ring for good performance at both the signal and oscillator frequencies. I published a design for a 5616-MHz LO multiplier chain in the May 1993 issue of *QEX* that will easily drive these diode mixers. You feed in the output of a 561.6-MHz no-tune LO and the chain puts out +7 or +15 dBm, depending on which output option you choose. I typically use +9 or +10 dBm to drive a pair of mixers with HSMS-8202 diodes. Using a higher IF for better image rejection often makes it necessary to compromise between good LO rejection and low conversion loss, since the $^6/_4$-wavelength ring is frequency sensitive.

While the Hewlett-Packard HSMS-2822 diodes sort of work, I don't recommend them if you can get HSMS-8202 Ku-band diodes. I get 1.5 to 3.8 dB less loss with the Ku-band diodes. More important, it wasn't necessary to tune the mixer to get good performance. One mixer I built showed only 5.3 dB of conversion loss without tuning. The input 1-dB compression point was +1.3 dBm. If you have a spectrum analyzer, one way to evaluate the diodes is to compare the upper and lower mixing sidebands of an untuned and unfiltered mixer. With good diodes, the two should be nearly identical, while diodes with excess stray reactances may result in significant differences of a dB or two in conversion loss. Of course, if you don't mind tuning the mixer with bits of copper foil, the 2822s are useable even at 10 GHz. I'd use + 13 to +15 dBm of LO drive with the 2822s.

The radial stubs for the mixer and splitter are of different sizes. I made the one for the splitter smaller to compensate for the lead inductance of real resistors. At 6 cm, even tiny chip resistors can have a significant amount of stray inductance.

It may be worthwhile to add quarter-wave RF chokes between the splitter outputs and ground. 400 mils of 30-gauge wire should be about right for 5616 MHz. This will isolate the mixers from each other at the IF. Of course, the lack of isolation could make testing easier, since you could use the transmit IF signal to tune up the receive filter with a spectrum analyzer. On the other hand, noise from the transmit IF circuitry could prevent your receiver from obtaining the expected low noise figure. In one 10-GHz transverter I built, this noise doubled the NF from 1 to 2 dB.

Mixer Construction

After etching the board on 15-mil 5880 Duroid, I trim the board with a shear into a rectangle, leaving copper foil out to the edges. This allows me to solder 0.50×0.025-inch brass strips around the mixer board to form a frame suitable for adding a cover. These strips are drilled and tapped for five SMA connectors. SMA connectors are overkill for the IF connection, but they result in a compact and RF-tight assembly. By using 2-hole flange connectors for the IF, you can offset the center pins so they clear the ground foils. I've also used Teflon feed-throughs, but these compromise the RF integrity of the assembly. It is possible to improve performance slightly by tuning the mixers with small pieces of copper foil, but this shouldn't be necessary. My May/June 1993 *QST* 10-GHz article discusses mixer tuning.

5760-MHz Band-Pass Filter

The low IF used by amateurs makes filtering a challenge. Often, amateurs resort to waveguide filters to generate a clean signal. This works, but 6-cm waveguide is a little big for my taste. (Plus, it isn't the most commonly found

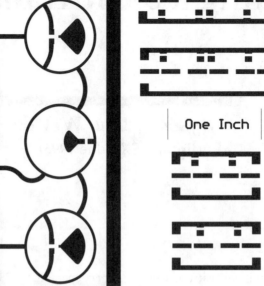

One Inch

Fig 1—Etching pattern of mixer and MMIC amplifier boards. Board material is 15-mil Rogers RT/Duroid 5880 with a dielectric constant of 2.2.

stuff, particularly at New England flea markets. I think it's around, but who wants to cart around heavy metal objects that few people want?) A pipe-cap filter is pretty easy, but as some amateurs have discovered, a single pipe-cap filter gives marginal performance, especially if you are fortunate enough to find a surplus amplifier capable of generating significant power.

By carefully evaluating the plots by Kent Britain in the 1988 *Microwave Update* and making an educated guess using my knowledge of filters, I came up with a two-resonator filter that seems rather easy to duplicate—if you have the proper equipment to tune it up. I tried improving on my initial guess by varying the probes, and small changes weren't too critical, particularly when compared to trying to adjust a single resonator design for lots of spurious attenuation. The design shown in Fig 1 has a bandwidth of 47 MHz with 3 dB of loss. Using a 145-MHz IF, unwanted mixing products are at least 43 dB down. This should be adequate if you are using low-side injection with typical surplus amplifiers.

Like most of you, I don't have a network analyzer to use to sweep the filters. Instead, I used a mixer/LO to upconvert a VHF/UHF signal generator. By terminating the mixer with an isolator, I get measurements that seem to make sense. I've tried in the past to make measurements with a multiplier/filter driven by the signal generator, but this required much more work and yielded rather fuzzy results. With the mixer, I was able to cover 400 MHz around 5760 with less than a dB of variation in spectrum analyzer response; it wasn't even necessary to use the correction factor needed with the multiplier setup.

The advantage of using more resonators is the resulting steeper skirts that better attenuate the unwanted mixing products—I found the Q of the resonators I built wasn't high enough to clean up the signal with just a single resonator. Using corrosion-resistant nickel-plated brass screws is part of the reason, but I don't have a source for small quantities of silver-plated screws of the right size. Brass and nickel are significantly less conductive than copper and silver, so losses are noticeably higher if you use tuning screws made of these materials. By the time I narrowed the bandwidth of the filter enough to get lots of stop-band attenuation, the losses became excessive. In addition, with a really narrow filter you need to start worrying about effects like mechanical and temperature stability. It isn't hard to see how a filter with flexible Teflon board supporting some probe pins could be easily detuned.

Filter Construction Details

The filters use a pair of 3/4-inch copper pipe caps. The one I measured actually had an inside diameter of 0.88 inches. They are tapped 8-32 at the top, though copper isn't the best material for holding threads. (You could solder brass nuts to the top, but then you would need longer screws than the 8-32 × 5/8-inch screws I used.) I estimate that the screws extend about 0.35 inches into the cavity. Polishing up the inside of the caps is a good idea—a smooth surface helps raise the Q of the resonators.

I've found that tightening the lock nut pulls out the screw slightly, raising the resonant frequency. Conversely, tightening the screw against the copper threads of the pipe cap moves the screw in a little, lowering the resonant frequency. Done just right, you can end up with a precisely tuned resonator with the lock nut at just the right tension. This is how I normally tune up my waveguides and cavities. Alternately, you might consider a better tuning mechanism that doesn't require as much skill. I saw an interesting Gunn oscillator design that used dielectric rods attached to piston tuning assemblies.

The resonators are coupled together with a 2.0-inch length of 0.085-inch diameter semi-rigid coax. While I didn't experiment with different lengths, I recommend you stick with this length of coax. 250 mils of shield are removed from each end of the coax, so that the dielectric is exposed. The dielectric is left on to protect the center conductor. It probably makes sense to bend the coax first, then drill the coupling holes for the resonators through the mounting board. You can then vary the 0.75-inch spacing to suit the coax, as it isn't critical. This may be easier than precisely bending the coax to match the spacing. The 0.500-inch spacing between the probes in the resonators *should* be maintained, unless you want to experiment with a new design. I made the mounting board out of unetched 1/16-inch-thick double-sided circuit board. The poor thermal conductivity of the fiberglass is an asset—you can solder the pipe caps to the board without unsoldering the probes. Including the #33 mounting holes is a good idea even if you don't intend to use them immediately, as adding them later might take a bit more work.

5760-MHz Amplifiers

Transmit amplifier response plays a big part in the cleanliness of your microwave signal. Most surplus amplifiers useable for 5.76-GHz amateur work are designed for operation at 5.9 to 6.4 GHz. Not surprisingly, those using high-side local oscillators require more filtering, since the amplifiers are actually optimized for the LO frequency. This often isn't a problem with retuned or homebrew amplifiers, since the tuning typically results in a narrowband amplifier with rejection off the tuned frequency. But, many surplus amps do work reasonably well in the amateur band without any tuning, so many people do use them "as is."

The new Mini-Circuits ERA MMICs are just what we need to take the filtered signal and amplify it up to a level adequate for driving TWTAs or surplus amplifiers. The cascade of ERA-2/3 MMICs shown in Fig 2 has 26 dB of gain and +14 dBm of linear output. To prevent unwanted feedback, the amplifiers are only 0.50 inches wide. This results in a cutoff frequency of 11.8 GHz—high enough to offer significant attenuation over the bandwidth of the MMICs. A much wider enclosure invites waveguide propagation unless hard-to-find microwave absorber material is used. The simple MMIC circuitry makes this easy to accomplish. Amplifiers using GaAsFETs are often much wider, in order to accommodate the low-loss bias circuitry. I've also included the etching pattern for an amplifier using just a single MMIC, for applications that don't require the gain of two MMICs.

MMIC Amplifier Construction Details

After etching the board on 15-mil 5880 Duroid (ε_r=2.2), I trimmed the board to 0.50 × 1.45 inches. Next, I drilled holes for the power leads and carefully cleared away the copper ground plane around these holes with a large drill

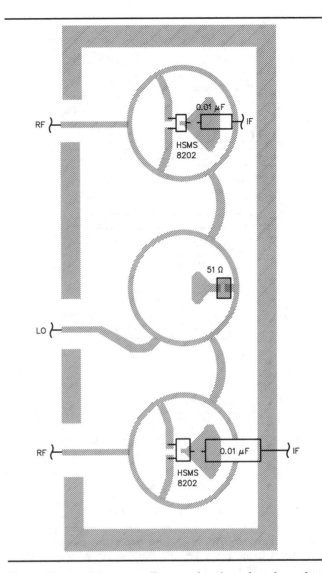

Fig 2—Parts-placement diagram for the mixer board.

bit. Practicing with some scrap Duroid and different drill bits is highly recommended—you don't want the bit to "grab" and ruin the board. I use dial calipers for laying out the brass strips that form the frame around the board, just as with the mixer board. The frame also holds the connectors and feedthrough capacitor. I ended up mounting the capacitor on the side of the box since there was no room to mount it on the output end plate. There is a convention that the dc input connector ought to be mounted next to the output connector, but there are numerous exceptions to this rule.

I punch a 94-mil hole for each of the MMICs to sit in, then bend the grounding leads flat against the MMIC and stick them through one of the holes. After the input and output leads are flush against the board, I bend the grounding leads flush against the copper ground plane and solder them down. Then I attach the other surface-mounted parts. This differs from the usual practice of mounting the semi-conductors last. I do it this way to ensure the best possible ground lead connections, which is critical for proper microwave performance. Finally, I wire up the regulator on the

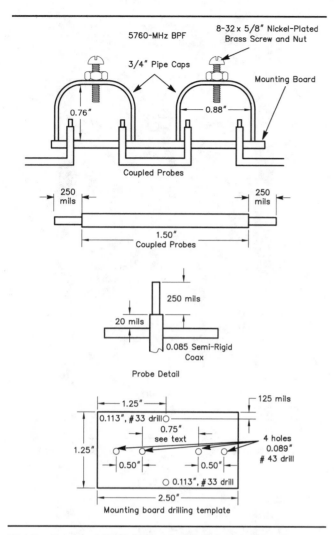

Fig 3—Coupled 5760-MHz pipe-cap band-pass filter. This filter provides a 47-MHz bandwidth with 3 dB of insertion loss.

ground plane side of the board.

There are two choices for a high-quality LNA on this band. The first is the design by Ranier, DJ9BV, in the March 1996 *QEX*. The second is the one I published in the September 1994 *QEX*. Ranier's has a slightly lower noise figure, but the 13 dB of typical gain isn't quite enough to overcome the noise generated by most 6-cm mixers, so a second stage will be required. My two-stage design with 22 dB of gain is about right for a terrestrial station. An EME station free of interference might effectively use as much as 30 dB of preamp gain ahead of the mixer to maintain a low system noise figure.

System Integration

Many of my transverters successfully use the Chip Angle sequencing circuit found in the *ARRL Handbook*. (Chapter 22, "T/R Time Delay Generator," by Chip Angle, N6CA.) As the diagram in Fig 6 shows, I avoid possible relay damage by sequencing both the RF drive to the transmit mixer and the dc power to the transmit amplifiers. While you can still end up hot-switching the relay if the relay sticks and then releases at the wrong time, I think this is so rare that I haven't bothered to design a suitable interlock. (I suppose you could sense the dc continuity of the relay and use this to indicate relay closure.) The September 1995 *QEX* has schematics that show the sequencer and IF circuitry in more detail.

I've found that this technique works just fine with semi-break-in transceivers since the transverter doesn't really care what the transceiver is doing (except for the dc control signal, obviously). Even if the transceiver is transmitting when the converter is receiving, everything is still operating acceptably—the 14-dB pad and switching-diode loss protects the receive MMIC from excessive RF. Thus, it even makes sense to use a center-off toggle switch to provide a manual

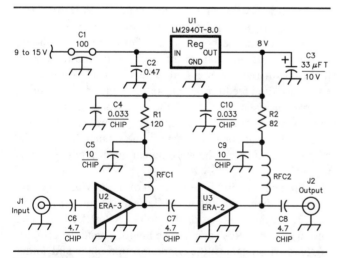

Fig 4—Schematic diagram for the MMIC amplifier.
RFC1, RFC2—3 turns no. 28 enam wire closewound, 0.062-inch inside diameter.
U1—National 2940T-8.0 low-dropout regulator.
U2—ERA-3 Mini-Circuits MMIC.
U3—ERA-2 Mini-Circuits MMIC

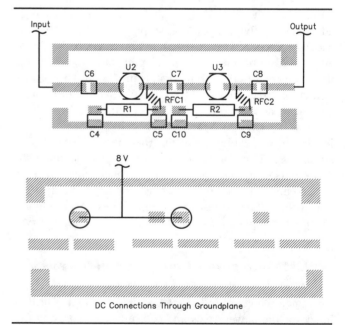

Fig 5—Parts-placement diagram for the MMIC amplifier.

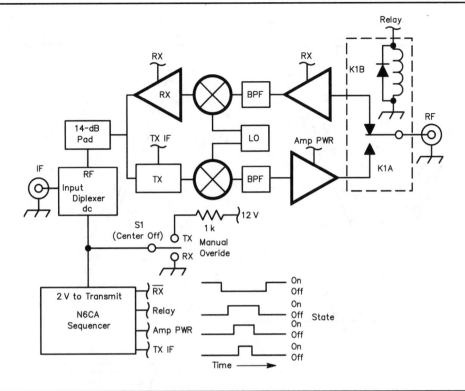

Fig 6—Using the N6CA sequencer board with a transverter and T/R relay.

override. This allows you to plug in any low-power IF radio and manually force the transverter into the proper state.

Those of us who travel 100 miles from home to operate from a tall mountain appreciate this feature. This is also handy for testing the transverter with test equipment, though there is a potential for an unexpected problem. Some signal generators cut back the RF if they see dc voltage on the output connector, so you might think there is a problem with the transmit converter when there really isn't; the transmit converter isn't putting out power simply because it isn't seeing much drive.

This scheme works well with low-power solid-state amplifiers because they are easy to turn on and off with power FETs. You can buy P-channel FETs with relatively low on-resistance—with even better devices appearing as time goes on. A simpler approach is to use the on/off control pin of the National LM2941 adjustable regulator, which offers a low drop-out voltage and delivers an amp. I'm not sure how well it will work with TWTAs, which often have warm up times and can't be turned on and off instantly, at least with inexpensive techniques.

Much less critical is the receive sequencing. Usually, the delay is so long that it doesn't matter whether the receive converter is turned on before or after the transmitter! In checking my previous work, it looks like I've done it successfully both ways. Turning the receive converter off first makes sense if you want to conserve as much current as possible, a consideration when running from batteries. I've not done any testing to see whether preamps are more resistant to high levels of RF with or without power applied, but the answer may determine the best approach with marginal relay isolation.

Connecting the Blocks

The coax and waveguide techniques that work well at 10 GHz work even better at 6 GHz, so most people don't have any problems. However, building a 10-GHz station isn't really a prerequisite to getting on 10 GHz, though almost everyone on 6 GHz also has 10 GHz. Finding 6-GHz surplus equipment is a lot easier than finding stuff suitable for narrowband 10 GHz, though sources do seem to be drying up.

At 6 GHz, you can pretty much toss out the idea of short pigtails; they just don't work with commonly found coaxial cables. You might be able to make the technique work with tiny 0.035 or 0.047-inch diameter semi-rigid coax, but I've never seen the stuff available inexpensively. By soldering the shield directly to the ground plane, you can get very short connections. The solid shield also helps, since you don't have to worry as much about stray wires shorting out the cable. Even at 2.3 GHz, pigtails seem to work only for low-power circuitry; I've had little success getting them to work above plastic MMIC power levels.

Normally, I connect all the assemblies together with SMA connectors and use N connectors for the antenna hardware. Testing cables and adapters at RF is a good idea—there is cheap hardware around that won't work well at this frequency. I've joked that one particularly cheap adapter could be used as an image stripping filter, due to its rather pronounced notch response. While I've left the BNC con-

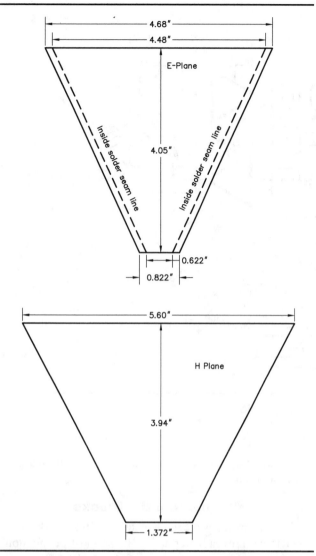

Fig 7—15.8-dBi 5760-MHz horn dimensions. The horn solders to WR-137 waveguide. The dotted line marks the inside solder seam.

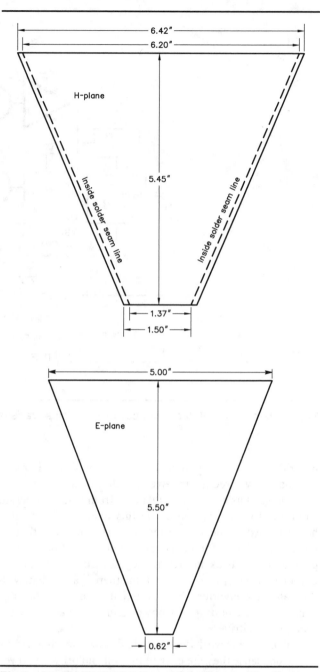

Fig 8—17-dBi 5760-MHz horn dimensions. The horn solders to WR-137 waveguide. The dotted line marks the inside solder seam.

nectors on our spectrum analyzer when doing quick checks, I don't recommend using them at 6 GHz.

At 6 GHz, your coaxial cable can actually be too big. If it supports more than one propagation mode, you can have signals propagated by different modes canceling each other out, resulting in extremely high losses. Andrew Corp suggests 5.0 GHz as the maximum frequency for LDF5-50A ⅞-inch Heliax. Of course, if you sweep the cable, you'll probably find frequencies above 5 GHz at which the cable works just fine. This is why many people are able to use ½-inch Heliax at 10.368 GHz: the frequency is in one of the clear windows above the cable's single-mode limit. I've had no trouble using either RG-213/U or ½-inch Superflex on 6 cm, although RG-213/U is rather lossy.

Horn Antennas

I included the simple antennas of Figs 7 and 8 to show an easy technique for mounting a small horn to a mast. Many textbooks and articles aren't clear how you accomplish this

task. I use pieces of waveguide that come with unusual flanges. The flanges provide rugged attachment points for screws. To me, this is a lot easier than trying to fabricate some sort of bracket or welding the horn to a suitable mounting plate. This also works well for slot antennas. Besides, I have no other use for the mounting flanges. The horns have predicted gains of 15.8 and 17.0 dBi, respectively.

I made the phase centers equal by making the horn H-plane width smaller while keeping the horn length and E-plane widths constant. Having them equal is useful if you intend to feed a lens. While you could make the E-plane width wider, you actually lose a little bit of gain, even though the horn is bigger. You could say that the transition to free space is occurring too quickly for maximum gain. This is why horn gains above 23 dBi are rare—dish antennas become much more practical than a very long horn. With the 17 dBi horn, I also made the horn a little longer.

Horn Construction

I made the smaller horn out of $^1/_{16}$-inch unetched double-sided fiberglass circuit board. For light weight, I made the 17-dBi horn of thinner 0.025-inch G-10 circuit board, though it's not as strong. For durability, I find it important to tape the joints with copper foil. Otherwise, people borrowing the horns return them with broken solder joints. To protect the copper from corrosion, I painted them with clear acrylic spray paint.

Since most people use coax on this band, Fig 9 shows an N-to-WR-137 transition. I used an Amphenol 82-97 UG-58A/U connector. These connectors have a center contact that press fits into the Teflon. A different connector may require a bit of experimentation, since the center contact forms part of the probe. Bare #12 copper wire is easily obtained by stripping ordinary house wire sold in hardware stores. I used a hacksaw to cut a pair of slots for the brass shorting plate. The 0.256-inch dimension is from the center of the probe to the inside surface of the shorting plate. I slid the snugly fitting 1.5×0.622×0.032-inch shorting plate through the slots and soldered it in place with a propane torch.

I soldered the horn directly to the waveguide with copper tape. A soldering iron is useful for tacking the tape into position. Then I used a propane torch to do the final soldering, since the waveguide needs quite a bit of heat to properly melt the solder. The small horn and large horns measured 14 dB and 30 dB return loss, respectively. I consider 14 dB adequate, though a purist would add tuning screws or vary the probe length for a better impedance match. People nor-

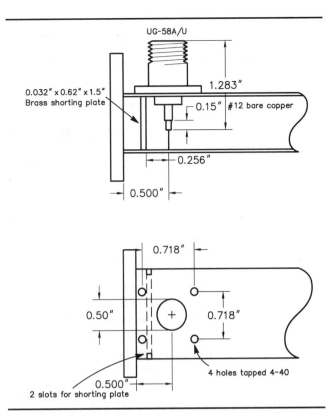

Fig 9—WR-137-to-N connector transition.

mally put screws on the centerline of the waveguide, where they have the most effect. Usually, two or three screws placed a quarter or eighth of a waveguide wavelength apart will match nearly any load. I wouldn't worry too much about not having a set-up to measure SWR—I made a couple of 200-mile 10-GHz contacts before I finally got a precision directional coupler and reduced the SWR of my dish feed below 2:1.

3456-MHz Transverter

By Zack Lau, W1VT

(From *QEX*, September 1996)

Jim Davey's 3456 no-tune transverter was an amazing piece of work—it was the first of the no-tunes and did a remarkable job of establishing the concept. Like many ground-breaking designs, it had a number of minor problems. Here is my approach to fixing them.

The most serious problem is board radiation that unbalances the mixers and degrades spectral purity. This is caused by the combination of a low dielectric constant and a relatively thick circuit board. Unfortunately, high dielectric constant boards have about twice as much loss, a significant drawback since the transverter requires high-Q band-pass filters. As a result, I chose a much thinner, 15-mil board. This allows an aluminum cover to be placed over the circuit with negligible effects on circuit performance. You don't need absorptive rubber to shield this circuit. Microwave absorber material can be tough to find in small quantities.

Another drawback of the original design is the lack of voltage regulators. Performance was seriously degraded as the batteries ran down. By using three-terminal regulators, the circuits work just fine between 9 and 15 V. The LM2940T-8.0 is shown in Fig 1; it not only features a low drop-out voltage, but offers reverse polarity protection in case you hook up your batteries backwards! Beware, you can turn the regulator into an oscillator by substituting an inadequate value for C9. You need a sufficiently large amount of high-quality capacitance for stability. For details, consider downloading the data sheet from National's WWW page: **http://www.national.com/pf/LMLM2940. html**.

Finally, the MMICs used were a bit marginal, operating near their upper frequency limit. This version uses newer MMICs with significantly enhanced performance at 3.5 GHz. The new MMICs have so much gain that it made sense to revise the circuit topology. Instead of dual mixers, I chose to use a single mixer and a splitter. It is common practice when using a single mixer to use the same bandpass filter for transmit and receive. It is placed between the splitter and the mixer. Terminating the mixer with the splitter improves performance, while adding little extra complexity to the circuit. The situation would be different with pipe-cap or waveguide filters—then I'd have to spend nearly twice as much time fabricating the filters. In this situation, board space is saved because I don't need to place a small rectangle between two large circles.

There is also a subtle advantage for those of us using

3456-MHz no-tune transceiver, 14 dBm output, 1.7 dB NF.

3456-MHz no-tune transceiver voltage regulators.

heat-transfer techniques for fabricating the boards. All the high-tolerance filters are concentrated in a relatively small area of the board. This makes it significantly easier to get an accurate reproduction. I noticed this when examining an LO board with low output (+4 dBm instead of +8 dBm). The output filter was stretched so much that the gaps between the resonators were a couple of mils wider than the design called for. This effectively raised the center frequency of the filters. As Table 1 indicates, you can still get useable performance with 3 dBm of LO drive. To assist you in fabricating the filters, the dimensions are shown in Fig 2.

The single mixer significantly reduces the circuit board area required. Only 16 square inches are needed for the LO multiplier and main transverter board, about ⅔ of the original. It also simplifies transmit/receive switching—the mixer can be hooked up directly to a +3 dBm VHF transceiver, like the Rick Campbell mini R2/T2/LM2.[1] The effect of the splitter loss on performance is negligible—the receiver still has

[1]Notes appear at the end of this section.

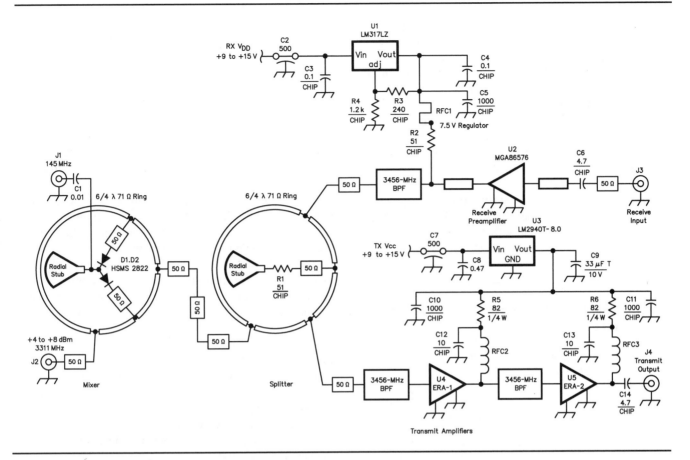

Fig 1—Schematic diagram of the 3456 transverter board.

C2, C7—Feedthrough capacitor; value not critical.
C6, C14—High-quality 4.7-pF chip capacitor like the ATC 100A. Not critical if you aren't worried about noise figure or power output.
C9—33-μF, 10-V tantalum. *The National Data Book* recommends a minimum of 22 μF.
D1, D2—HSMS 2822 packaged diode pair.
J1—Two-hole flange-mount SMA panel jack. Omni Spectra 2052-1652-02 works quite well.

J2, J3—Four-hole flange-mount SMA panel jack.
RFC1—Printed circuit board RF choke.
RFC2, RFC3—4 turns of no. 28 enameled wire close wound. 0.062-inch inside diameter.
U1—TO-92 case adjustable regulator.
U2—Hewlett-Packard MGA 86576 GaAs MMIC.
U3—National LM2940T-8.0 low-drop-out regulator.
U4—Mini-Circuits ERA-1 HBT MMIC.
U5—Mini-Circuits ERA-2 HBT MMIC.

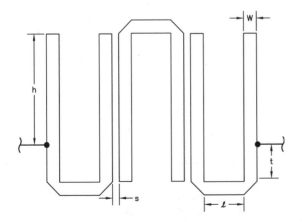

Fig 2—Dimensions of the 3312 and 3456-MHz band-pass filters on 15-mil 5880 Rogers Duroid.

		3456 MHz	*3312 MHz*
h	coupled line height (mils)	486	508
s	spacing between coupled lines (mils)	25*	25*
t	tap height (mils)	28	34
w	line width (mils)	50	50
ℓ	uncoupled length (mils)	150	150

*Modeled spacing, see text.

a 1.7-dB NF and 14 dB of gain, while the transmitter has a 1-dB compression point of +14 dBm with 12 dB of gain. The decrease in gain on receive may actually be an advantage by reducing its susceptibility to mixer overload if a low-noise preamplifier is added. The output level is convenient for running surplus TWTAs or their solid-state replacements.

The new Mini-Circuits ERA-1 and ERA-2 have just the right amount of gain for the transmit amplifiers. On transmit, too much gain can be just as bad as too little gain. The more gain you have, the easier it is to make an oscillator. Feedback might help, but installing feedback networks deviates from the idea of a simple, reliable project with a minimum of parts. These MMICs use heterojunction-bipolar-transistor (HBT) technology. They do a good job of combining wide bandwidth with a relatively low supply voltage. At 50 mA each, they draw a fair amount of current

to generate 20 to 40 mW of RF. I think this is a reasonable trade-off, considering the complexity of the alternatives. A discrete FET design would be more complex, requiring a lot more design work.

I haven't experimented with the new ERA-4 or ERA-5 MMICs to see if more output power can be obtained without modifying the board. I'm still waiting for the ones I ordered in late May '96. I don't recommend using the high-gain ERA-3 MMIC between the band-pass filters—it is quite likely to be unstable unless drastic measures are taken. It may be necessary to shield the printed circuit board filters from each other. You might also experiment with narrowing the "waveguide" enclosing the circuitry. You may significantly reduce the chance of unwanted waveguide propagation by installing a shield on the optional grounding strip in the center of the board. This shield would lower the cut-off frequency by a factor of two. The low-loss nature of waveguide is a significant disadvantage when attempting to build amplifiers—what better way to create an oscillator than to couple the input and output together with a low-loss transmission line?

Keep the parts close to the board to reduce their ability to launch signals into the waveguide. I laid out the board placing the RF chokes close to the edge of the waveguide, as opposed to the center. Objects in the center of the waveguide couple into the waveguide better than those close to the edges. This is why you typically put detector diodes in the center of the waveguide when you want to maximize the signal to the diodes.

Keeping to the idea of simplicity, the receive preamplifier is a single Hewlett-Packard MGA 86576 GaAs MMIC. It has about 24 dB of gain and a 1.6-dB noise figure. The NF

Table 1
Effect of local oscillator power on transverter performance

| Transmitter | | Receiver | |
LO Power (dBm)	Output Power (dBm)	NF (dB)	Gain (dB)
3.2	13.27	1.79	15.11
3.8	13.33	1.78	15.15
4.9	13.40	1.77	15.28
5.6	13.42	1.79	15.30
6.0	13.37	1.79	15.34
6.8	13.37	1.76	15.34
7.2	13.40	1.78	15.33
7.7	13.38	1.77	15.35
7.9	13.38	1.77	15.36

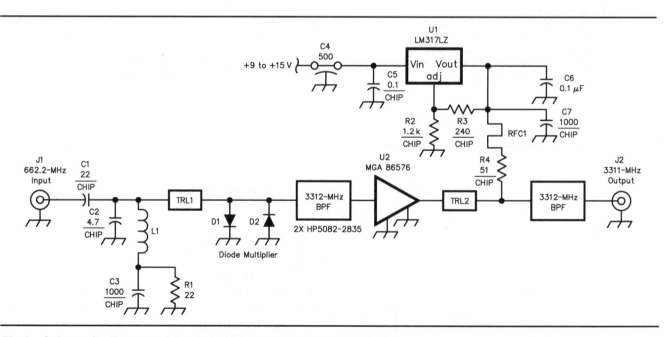

Fig 3—Schematic diagram of the 3311-MHz local oscillator multiplier.

C4—Feedthrough capacitor, value not critical.
D1, D2—Hewlett Packard 5082-2835 Schottky diodes.
L1—3 turns no. 28 enameled wire spaced 2 wire

diameters. 0.089-inch inside diameter.
RFC1—Printed circuit board inductor.
U1—LM317LZ adjustable regulator.
U2—Hewlett-Packard MGA 86576 GaAs MMIC.

is degraded an additional 0.1 dB by the 10 dB of converter losses. The LM317L could be replaced by a 78L07 or 78L06 regulator, but these are harder to find and require a larger input bypass capacitor for stability. With the three MMICs I used, the highest supply voltage resulted in the best gain and noise-figure performance. I didn't do any testing past the recommended 7-V device voltage. Unit 2 still had a 1.9-dB NF and 13.65 dB of conversion gain with a device voltage of 4.82 V, so the device voltage isn't terribly critical.

The mixer and splitter both use a 180° hybrid. A good reference on these may be found in Chapter 6 of the *ARRL UHF/Microwave Experimenter's Manual*. The narrow bandwidth of the hybrid isn't a problem in this mixer application, due to the relatively low IF of 145 MHz. This is only 4% of the center frequency. The radial stub for the splitter is a little small, in an attempt to compensate for the stray inductance of the chip resistor.

The LO multiplier uses a diode multiplier, a pair of band-pass filters, and an MGA 86576 GaAs MMIC. FETs can be more efficient than diodes at frequency multiplication, but they tend to be more critical with regard to drive and tuning. The GaAs MMIC is a bit more expensive than a pair of the new ERA-3 MMICs, however, they produce just the right power level for driving a single mixer. An ERA-3 is more appropriate for driving a pair of diode mixers. Another advantage of the MGA 86576 is that it draws only 16 mA, compared to 35 mA for a single ERA-3. Actually, a pair of ERA-3s has too much gain, and running an ERA-1 and ERA-2 in cascade ups the current draw to 100 mA. This is six times as much as the GaAs MMIC draws.

The LO multiplier is designed to be used with a 662.2-MHz source. It can be easily modified to work with a more standard 552-MHz source by removing either one of the multiplier diodes, D1 or D2. As a 6× multiplier, +15 to +20 dBm of drive is needed. As a 5× multiplier, +13 to +20 dBm of drive can be used. This multiplier works much better with the correct number of diodes. If the stability of the LO amplifiers is marginal, it may be wise to add a 100-Ω resistor or resistive pad to the input of the multiplier. It isn't too difficult to envision cases where the constantly changing impedance of the diodes could cause problems.

I know that a lot of people are looking for practical ways to design the microstrip hairpin filters. The simple answer is there isn't any, at least for amateurs with little time or money. The programs that accurately simulate the discontinuities, such as the bends in the microstrip, still cost quite a bit of money. Trial and error, particularly with return loss measurements, is an effective way of designing complex filters, if you have the time. If you have a spectrum analyzer, I've found that upconverting a low-frequency signal generator with a mixer makes a good signal generator—if you add an isolator. The isolator significantly reduces the interaction between the filter and the mixer. If you aren't careful, some mixer/filter combinations can actually indicate that the filters have gain. Reflected signals can actually enhance the desired signal. As Jim Davey found out, return loss is a much more sensitive indicator of circuit performance than insertion loss.

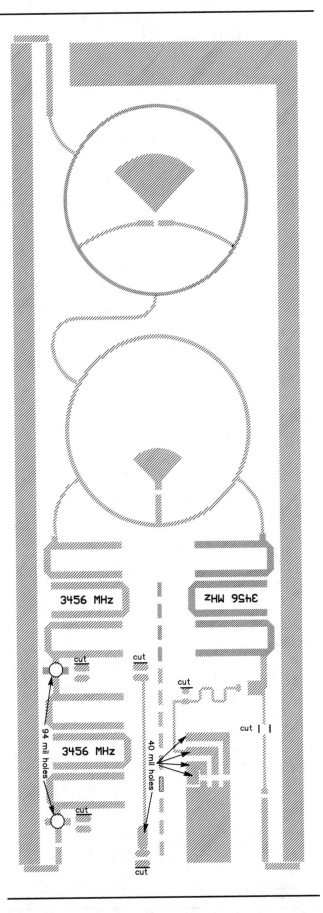

Fig 4—Cutting and drilling diagram for the transverter board.

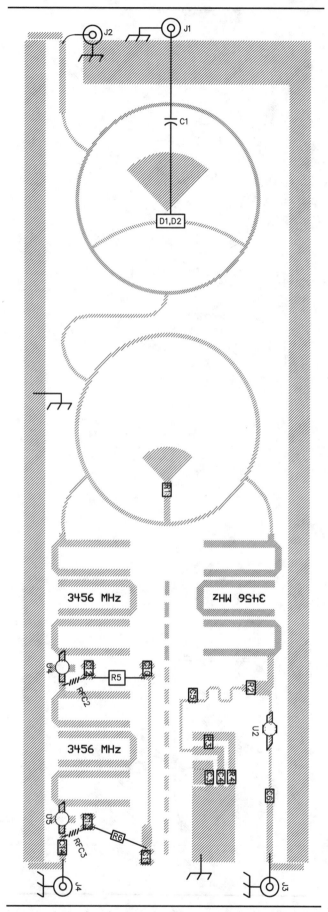

Fig 5—Trace side parts placement diagram for the transverter board.

Construction

Making these circuit boards is a stiff challenge—but some of us like challenges. Jim specified a tolerance of ±0.001 inches for the original boards. I doubt that I come that close with my etching techniques, but it's a target to shoot for. I was able to build three transverter boards, all with acceptable performance (Table 2). Mixing and LO spurs were at least 40 dB down. The second and third harmonics were only down 39 and 32 dB on one unit. This is no surprise, since there is no low-pass filtering of the output stage.

When comparing the dimensions of Fig 2 against the artwork, the careful examiner may notice that the spacing between coupled lines is actually 24 mils, as opposed to the 25 mils determined with the assistance of computer modeling. Since this was noted after several units were built and tested, I didn't bother revising the artwork. Due to the tight tolerances, I'd recommend you work with the Postscript files. If you can't download them off the 'Net from **http://www. arrl.org/qexfiles**, I can supply the file for noncommercial purposes if you enclose a 3.5-inch disk and addressed return envelope with postage.

The boards are etched on Rogers RT/Duroid 5880 with a dielectric constant of 2.20, clad with 1 ounce rolled copper on two sides. The dielectric thickness is 0.015 inches.

When trimming the circuit board, don't forget to leave enough room for the SMA connectors. Normal square-flange connectors are 0.5 inches wide—thus the board needs to be at least 0.25 inches from the center of the 50-Ω microstrip traces to the edge. On the other hand, excessively widening the board increases the possibility of waveguide propagation, so you don't want to err too far in the other direction either. You can also use smaller SMA connectors—Digi-Key now advertises a line by Johnson Components (formerly EF Johnson Components). While a bit expensive, they supposedly work up to 26.5 GHz, as opposed to 18 GHz for standard connectors. They should be useful for 24-GHz work.

I used a hobby knife with a new no. 11 blade to cut the slits in the board for the grounding straps and MGA MMIC ground leads. The blade of the knife should just touch the outer edges of the pads marking where to cut the slits. I don't trim away any Teflon from the slits. Instead, I use a flat-bladed screwdriver to carefully close up the holes after the leads are passed through by reworking the remaining copper foil. The holes for the ERA MMICs are punched with a

Table 2
Test data for three converters

| | Transmit | | Receive | |
| Unit | Power (dBm) | Gain (dB) | NF (dB) | Gain(dB) |
	(1 dB compression)		(compressed)	
1	14.40	12.9	1.78	15.34
2	14.10	12.6	1.75	14.95
3	14.83	12.1	1.64	12.94

Power was measured with an HP 8563E spectrum analyzer and confirmed with an HP 8481A/435B power meter. An HP 8970/346A was used to measure receive converter performance. Both an HP 8640B and a Marconi 2041 were used to generate the IF drive.

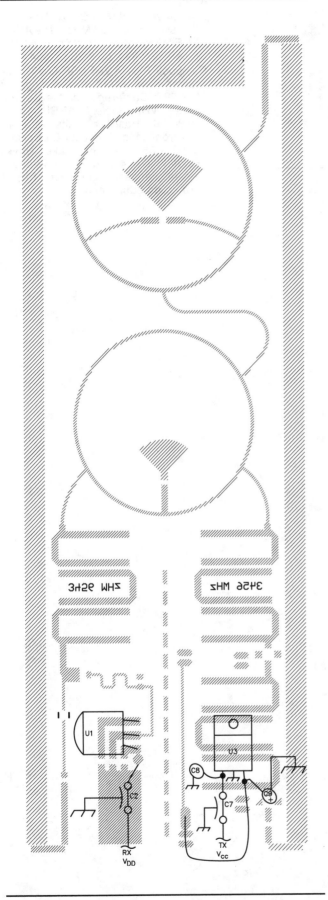

Fig 6—Ground-plane side parts placement diagram for the transverter board.

552 to 3312-MHz LO multiplier on 15-mil 5880 Duroid.

94-mil hole punch and then touched up with the hobby knife. I've found that drilling large clean holes in thin Teflon board can be difficult. After drilling the 40-mil holes for the power supply parts, I countersink the ground plane side by hand with a $1/4$-inch drill bit.

Some drill bits bite too deeply into the board, so you might practice your techniques on a few scraps of Teflon board. If you work slowly and realize there is a problem, you can "save" the board by carving away the excess copper with a sharp knife.

Even with a frame made of 25×500-mil brass sheet stock around the edges, I felt the board could use a bit of stiffening. To stiffen the board, I soldered some thin strips of unetched double sided circuit board to the ground-plane side. The assembly was stiffest when the boards were slid against the metal frame and soldered it. Forming an X or two seems to markedly stiffen the board. Brass could be used, but the circuit board is easier to solder and has less flexibility. A stiff board is important because most chip parts can't flex along with the board—too much flexing and something will break.

A two-hole flange mount works well for input connector J1. The thinner connector can be raised slightly compared to the other three connectors. This makes it easy to connect the input capacitor, C1. Alternately, higher walls could be used to enclose the transverter board.

While there are pads for grounding the ERA MMICs with copper foil, I found it easier just to bend and solder the leads. The leads are inserted into the holes and then bent against the body of the MMIC, flattening them against the ground plane. I used copper foil with Unit 2—there wasn't a significant difference in performance compared to the other two units.

I first test the voltage regulators to make sure they are putting out the proper voltages before installing the resistors that supply power to the MMICs. The actual voltage at the MMICs can vary a bit—the GaAs MMIC can draw between 9 and 22 mA, so the voltage drop across the 51-Ω chip resistor can be anywhere from 0.46 to 1.1 V. I'm not surprised at this range—manufacturing repeatable bias points has always been a weak point of FET technology compared to bipolar. The device voltage for the ERA MMICs is supposed to be between 3.2 and 4.4 V, nominally 3.8 V.

The shared mixer ought to make troubleshooting easier. If nothing works, there is a problem in the LO/mixer, or in both the transmit and receive amplifiers. While the conversion gain isn't excessive, it ought to be enough to hear the increase in noise when you turn on the receive amplifier.

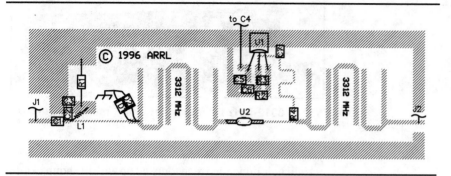

Fig 7—Parts placement diagram for the multiplier board.

Notes

[1]Jim Davey, WA8NLC "A No-Tune Transverter for 3456 MHz," *The ARRL UHF/Microwave Projects Manual*, Vol 1, ARRL, 1994, pp 3-28 to 3-34.

[2]Campbell, Rick "A VHF SSB-CW Transceiver with VXO," *Proceedings of the 29th Conference of the Central States VHF Society*, ARRL, 1995, pp 94-106. Boards and kits are available from Kanga Products, Bill Kelsey, N8ET, 3521 Spring Lake Drive, Findlay, OH 45840, 419-432-4604. e-mail: kanga@bright.net or **http:// qrp.cc.nd.edu/kanga/**.

2304 and 3456 No-Tune Transverter Updates

By Steve Kostro, N2CEI,
Down East Microwave, Inc.

(From *The Proceedings of 1996 Microwave Update*)

Since the introduction of the "NO-TUNE" transverters in the late 1980s[1] Monolithic Microwave Integrated Circuits, MMICs, have greatly improved in performance and yet remained economical enough for amateur use. This article will discuss various options and some results in improving the performance of the two transverters originally designed by Jim Davey, WA8LNC. The three basic issues that will be covered will be Receiver Performance and Options, Transmit Power and Purity, and Local Oscillator Multiplication. Both transverters will be discussed simultaneously allowing comparisons to be made in construction and end results with some words on construction technique. Since detailed drawings of component placement are included, it was not found necessary to include schematics.

Receiver Performance and Options

When the receiver sections of the two transverters were

[1] Notes appear at the end of this section.

originally designed, MMICs with noise figures of less than 4 dB were not economically available to most amateurs. Although low noise GaAsFET preamplifier designs[2] that were being used by hams were considered, they required special bias supplies and a large amount of circuit board space. With the scope of the project being simplicity and repeatability, and driven by economics, GaAsFETs were deemed not practical to implement into a simple transverter design. Today, with the advent of affordable GaAsFET MMICs, we now have stable drop-in, or with a little circuit modification, designs that will have less than a 2 dB NF and over 20 dB gain. The HP/Avantek MGA85676 became the MMIC of choice for both units (Figures 1 and 2). It replaces the first MAR 6 MMIC in both transverters and will produce a 2-3 dB NF improvement. (See Table 1 for approximate specifications of the MGA85676.) The transverter's noise figure could be optimized for <2 dB by performing the input circuit modification shown. With the increased gain that the MGA85676 offers, the proceeding MMICs should be com-

Table 1

MGA86576

Frequency	Gain	50 ohm NF	Optimized NF	I drain @ Op Volts
2 GHz	>22 dB	2.1 dB	1.5 dB	16 mA @ 5 V
4 GHz	>22 dB	2.0 dB	1.6 dB	16 mA @ 5 V

Table 2

Model #	Gain @ 2 GHz	Gain @ 4 GHz	Pout @ 1 dB comp	I drain @ Op Volts
ERA-1	>11 dB	>10 dB	+13 dBm @ 2 GHz	50 mA @ 3.8 V
ERA-2	>14 dB	>13 dB	+14 dBm @ 2 GHz	50 mA @ 3.8 V
ERA-3	>19 dB	>16 dB	+11 dBm @ 2 GHz	35 mA @ 3.8 V
ERA-4	>14 dB	>13 dB	+19 dBm @ 1 GHz	80 mA @ 5 V
ERA-5	>18 dB	>15 dB	+19 dBm @ 1 GHz	80 mA @ 5 V

The tables above are to be used as a guideline for Amateur Radio design work. They do not completely reflect the manufacturers' specifications. Please consult the manufacturer for complete data.

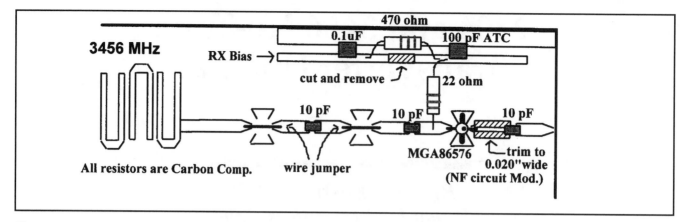

Figure 1

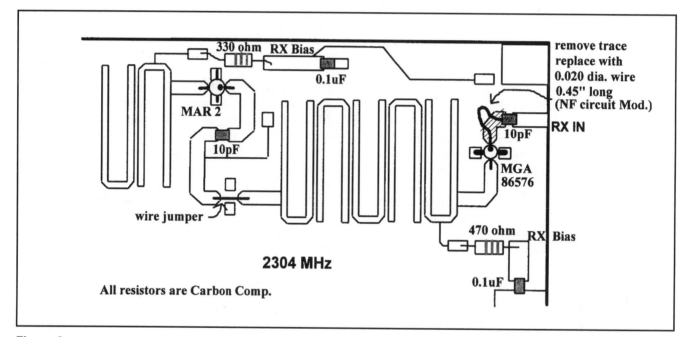

Figure 2

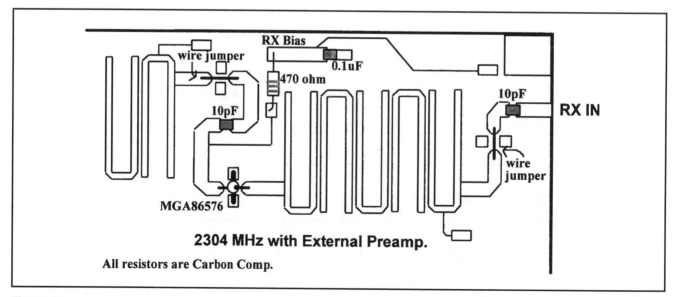

Figure 3

pletely removed on the 3456 transverter. On the 2304 unit, remove the second gain stage (between the filters) and replace the third stage with a MAR 2. If you plan to use an external GaAsFET LNA with the 2304 transverter (Figure 3), remove all three gain stages and install the MGA85676 MMIC in the second stage position, between the two bandpass filters, to ensure that the out of band products amplified by the preamp are filtered out. If an external LNA is to be used with the 3456 transverter (Figure 4), a filter should be installed (pipe-cap[3]) ahead of the transverter. It will not be necessary to perform the low noise figure modification on the MGA86576 if it is used as a second stage.

When implementing these changes, check all ground wraps and/or ground rivets. Be sure they are as flat as possible and are clean connections. The MGA86576 will oscillate if there is an excessive amount of inductance on its ground leads. Also verify that there is a good ground connection at every by-pass capacitor! If you are modifying an old board, oxidized, missing, or broken ground connections could have been the original problem keeping the transverter from working in the first place. Take your time in checking them out this time (It's a thought!).

Transmit Power and Purity

With the introduction of Mini Circuit Laboratory's ERA series of MMICs (Table 2), it is now possible to have as much as 100 mW of output power on both bands. Both transverters used a MAR-8 as a final power amplifier that at times because of its low frequency gain, became very unstable. If you managed to tame it down, 5 to 10 mW output was realistic. There were a few mods that incorporated a single gate GaAsFET as a replacement, but they were not a drop-in solution. The other problem with the TX circuit was that the radiated LO signals, fundamental and multiplied, would conduct into the final gain stage and became amplified to levels as much as –10 dBc. The radi-

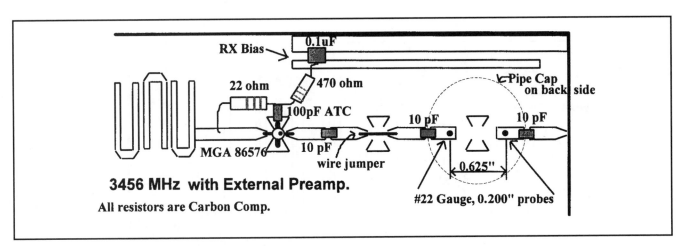

Figure 4

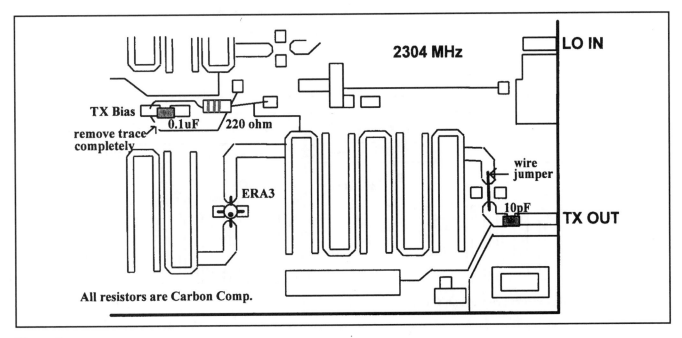

Figure 5

ated LO is a separate problem and will be addressed later in this article, but with the new MMICs and a brief explanation, both problems will be improved on.

The 2304 transverter uses two gain stages on the TX side, a MAR 3 and a MAR 8. Remove both of them. The 2304 PCB design (REV. D) only cycles the bias on the output MMIC. This needs to be changed so the TX bias is only on during TX for both stages. Perform the suggested modification to the circuit board in Figure 5. (Cut and remove the trace!) For a cleaner than original signal, install a ERA-3 in the position between the two filters . This is all that is needed for 10 to 15 mW of output. If higher output is desired, a ERA-4 could be installed in the final output position (Figure 6). If the LO is not modified, expect to see close to 100 mW output at 2304 and as much as 10 mW of output at 2160. This might be okay if you plan to use an extra band-pass filter on the output before amplifying it by any other gain stages. If other output levels are required, select any other MMIC that meets your specification.

The 3456 transverter is very similar to the 2304 unit but it used three gain stages in the TX chain. Remove all of them. In this design, you have a little more flexibility. If you figure a clean −15 dBm out of the mixer/filter combination, decide what you would want for an output, refer to Table 2 and do the math. Only recommendations are not to cascade any other MMIC with an ERA-3 between the two band-pass filters. A good combination (Figure 7) would be to use a ERA-2 driving a ERA-1 between the filters. This alone will produce 10 to 15 mW of output. If higher power is desired, an ERA-3 between the two filters driving an ERA-4 final (Figure 8) will deliver up to 100 mW output but will require a external band-pass filter if additional gain stages are to be used. Again it is recommended that the LO chain be modified to improve the spectral purity.

Like in the receive section, care must be taken in the assembly and to be sure about the grounding. If there are any doubts, fix it. Also if using the ERA-4s or 5s, care must be taken because they will dissipate a lot more power than

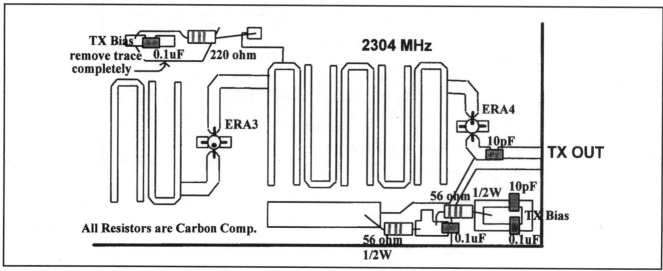

Figure 6

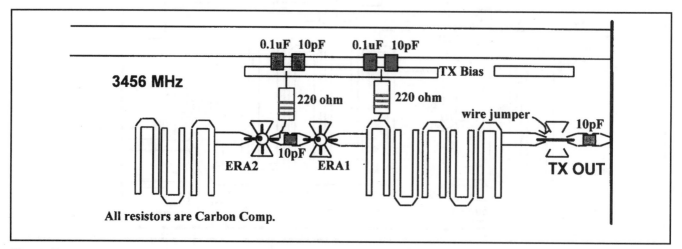

Figure 7

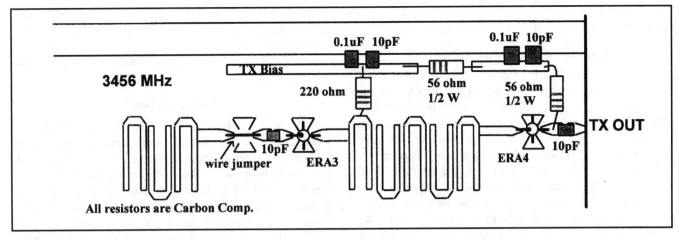

Figure 8

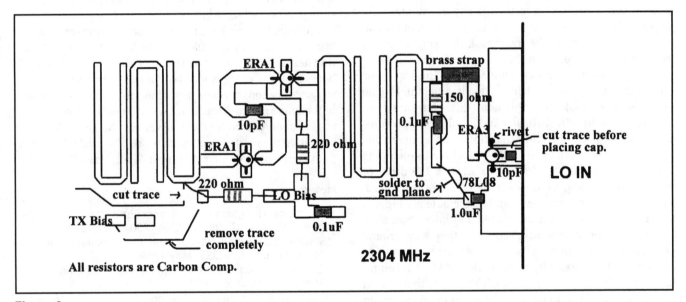

Figure 9

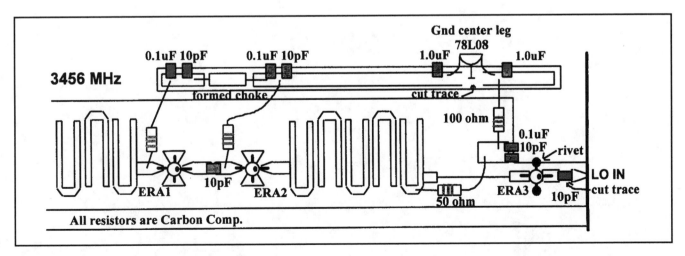

Figure 10

the older MMICs. Half-watt resistors get real warm! If you make a resistor change, do the math!

Local Oscillator Multiplication

The most difficulty encountered in using both transverters has always been caused by the 4X and 6X multiplier circuits. There have been many fixes and cures published and talked about, but the problems still remained. The major problem being the high level of 540 and 552 MHz. Energy injected into the harmonic generator circuit would be reflected back or radiated resulting in a starved amplifier chain, which would allow it to oscillate at its own desired frequency. With two fundamental oscillators now being introduced to the mixers, (the main and the self oscillator) it is easy to understand why the transverters worked sometimes really well and not so good all of the other times. Receiver noise is a good indication of operation but at what frequency would it be at if the LO chain had its own frequency being generated? Both transverters have fooled many noise figure meters. If any of the mods are performed on the two transverters, the LO chain modification would put you on the right track in fixing all of the transverter's other problems.

After reading an article published by Jim Davey on Frequency Multipliers[4] and discussing it with him, I found that it was possible to implement this idea into both transverters with little difficulty and produce huge results. Using the ERA-3 as the MMIC multiplier, it was found that a "sweet spot" existed as far as drive level versus operating voltage. A voltage was derived to allow a LO drive of +0 dBm with a plus or minus delta of 3 dB to be used. Remove all of the components and reassemble per Figures 9 and 10. When installing the ERA-3 multiplier, care must be taken in proper placement and grounding. New ground pads need to be formed with rivets. As for the amplifier section, all of the instability of the MAR-8 MMICs are eliminated along with the radiation effect. The new multiplier chain will deliver 10 mW of LO drive to the power dividers without even winding a coil and having nothing to tune.

The standard LO designed by Rick Campbell, KK7B,[5] will need the final MMIC removed from the circuit and replaced with a wire jumper. A +3 dBm signal should be the maximum obtainable level and will work correctly with the new multiplier circuits. Future discussion of the KK7B LO concerning frequency netting and stability will be discussed at a later date.

Final Notes

The modifications have been used repeatedly for over a year with excellent results. Local oscillators have been packaged together with the transverters that produce < 2 dB NF, 25 mW Pout and spurious responses of –45 dBc. Please remember that all of the new MMICs in the world don't make up for poor construction techniques. Use nothing but carbon composition resistors and keep the leads to the active circuits as short as possible. If you have a doubt about a ground wrap or rivet, repair it! Make sure that every by-pass capacitor is directly connected to a ground to the back side of the board. With the modifications, enclosures are less critical, but still could be a problem. Keep away from waveguide sizes boxes.

Contact the author at Down East Microwave (Phone: 908-996-3584, Fax: 908-946-3072, Web site: **http:// downeastmicrowave.com/**) for availability of the latest PC boards. Future designs will be available that use all of the suggestions above. Some new features of the new designs will include plated through holes and a machined enclosure to accommodate both new transverters. As newer MMICs appear on the marketplace, updates will be made. Feel free to experiment with them and most of all, Have Fun!

Notes

[1] *QST*, June 1989 for the 3456 transverter, *Microwave Update 1989* & *QST*, Dec 1992 for 2304 transverter
[2] "Simple Low-Noise Microwave Preamplifiers for 2.3 Through 10 GHz," *QST*, May 1989
[3] "Cheap Microwave Filters From Copper Plumbing Caps," *Microwave Update, 1988*
[4] *ARRL UHF/Microwave Projects Manual*, pages 5-13 through 5-15
[5] "A Clean, Low-Cost Microwave Local Oscillator," *QST*, July 1989

Modernizing The 3456 MHz No-Tune Transverter

By Jim Davey, WA8NLC

(From *The Proceedings of 1994 Microwave Update*)

Introduction

Back in 1988 I presented a single board no-tune transverter for 3456 MHz at the Microwave Update '88 Conference in Estes Park, CO. Initial interest was high in this design and it was subsequently published in the June, 1989 issue of *QST*. Since that time over 190 of the transverters have been sold all over the world.

While the 3456 MHz unit has enabled many to easily get a basic station on the band, it has not been without a few problems. Stability of the active devices has been marginal under certain conditions. Some people have also had a problem packaging the finished board in a larger enclosure with the other stages needed to make a high performance transverter.

Since the time of the initial design a couple of things have happened: (1) new and better MMICs have become available and (2) I have become more enlightened on the art of microwave circuit construction. Also, many people have had a chance to work out some of the problems, most notably WB5LUA, N1BWT and WA3JUF. Their willingness to take the time to publish their improvements is greatly appreciated. This paper will gather together the suggested modifications made by others and add some new ideas of my own.

Before getting into a discussion of the shortcomings and improvements of the transverter, I should reiterate the basic design goals of the unit:

1) All filtering is done by non-critical band-pass filters. This remains the dominant feature. No metalwork or filter tuning is required to achieve a reasonably clean transmit signal (unwanted outputs down >40 dBc) and an image-free receiver.

2) Low cost. If anything, silicon MMICs have gotten cheaper since the design was first published. And where can you get a decent 3456 MHz mixer for under a buck?

3) Integrated approach to multistage circuitry. Having all the essentials on one board reduces the amount of packaging, connectors and cabling required to get the unit together.

I believe that (1) and (2) above are still important features. The more hamfests I go to, the less I am inclined to emphasize (3), especially when I see boxfuls of SMA jumper cables at 50 cents apiece.

Packaging the Transverter

Let's face it, microstrip circuits radiate[1] and this transverter is no exception. The amount of radiation is high enough to cause some problems and has made it difficult to box up the transverter in a practical matter. Also, when an outboard preamp and/or power amp is desired, the isolation between those accessories and the board may not be high enough to allow them to be packaged together without feedback. Several people have tried to cover the suggested "brass strip box" in order to isolate the low level stages from the rest of the station. Generally this has caused problems ranging from high level spurs in the transmitter to oscillation of the gain stages.

The effect of an RF-tight lid on the transverter can be almost totally eliminated by using some true RF absorber material on the inside of the lid. I tried this experiment with some of the $^1/_8$-inch thick material available through Tom Hill[2], WA3RMX/7, and found it to do the trick. Black carbon foam does not work! Without the RF absorber, one of the most noticeable effects is that the metal cover will unbalance the mixer sending lots of local oscillator energy through the transmitter. The onboard filtering is not sufficient to eliminate the extra LO.

Another observable effect of board radiation is the coupling across the board from one circuit to another (LO multiplier to TX amp for instance) and along the length of the board (TX mixer to TX output). This can be eliminated by boxing the TX, LO and RX in separate enclosures of the proper size. I cut a transverter board into TX, LO and RX sub-modules using a sharp knife and connectorized all the ports. Each board ends up about 1.3 inches wide by 6 inches long. When boxed up using the suggested $^3/_4$-inch brass strip and a cover made of thin aluminum, the resulting enclosure looks like a waveguide-below-cutoff. Each cover still needs to be treated with RF absorber material as described above.

Unless the transverter is totally redesigned on much thinner board to reduce the radiation effects, the above suggestions, combined with those for stability described in the next section, can produce a stable and clean transverter.

Stability Improvements

Instability has been a sporadic problem in three areas:

1) The cascaded MSA 0685/0685/0185 in the receiver can oscillate depending on source SWR, lead dress of the biasing components and power supply voltage.

2) The cascaded MSA 0885s in the LO amplifier can oscillate at VHF frequencies due to excessive gain below 1 GHz.

3) The MSA 0885 transmitter output MMIC may be unstable into certain loads.

The first problem above was largely eliminated by replacing the original 5 pF series coupling capacitors in the receiver with 1.0 pF. This reduced the low frequency gain enough to keep things under control most of the time. All the Down East Microwave units use the smaller value capacitor. Varying power supply voltage (as when using batteries) can still change the device S-parameters enough to get an oscillating condition. Paul Wade, N1BWE, described using an LM 2941 CT low dropout regulator to eliminate this problem[3]. Down East Microwave sells these regulator ICs as well as a board level kit.

Another solution is now possible for only a few extra dollars by using one of the new Hewlett-Packard MGA 86576 GaAs MMICs. This device has a flat gain curve down to low VHF frequencies and a much lower intrinsic noise figure. It requires a +5 volt supply, thus a regulator from the normal 12 volt supply is suggested by HP. I put one in stage #3 of the receiver and cut off the first two stages before enclosing the board in a brass box (Figure 1). Not only did the system noise figure drop from 5 to 2.0, but the receiver now only requires 16 mA of current instead of the 45 mA with the silicon MMICs. Rus Healy, NJ2L, has done this modification to his rover station with similar results.

The potential solutions for eliminating oscillation in the LO multiplier amplifier are somewhat limited because the LO needs to generate enough power to drive two mixers. A possible alternative MMIC lineup that would meet the gain and output power requirements would be an MSA 0385/0385/0986. Gain at VHF would be reduced considerably over the 0885 pair, but there is not a lot of room on the existing board for three stages and their bias resistors, and the three stage amp would draw over 100 mA. I did not try this modification. Instead, I took a look again at what the MGA 86576 could do. The specs say it cannot produce more than about +7 dBm at 3312, so the mixers would have a little less injection than before after filtering and splitting. To see what could be done I tried a single MGA 86576 at 5 volts. The circuit is the same as Figure 1 except the 5 pF input capacitor is not needed. The power to each mixer measured +4 dBm, which was about 2 dB more than expected. The output spectrum was exceptionally clean with

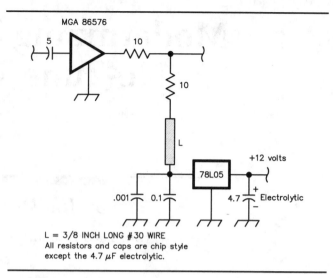

Figure 1

all unwanted products down over 70 dB. As a bonus, I found that the diode multiplier only required +8.5 dBm at 552 to saturate the MGA 86576, about a 7 dB reduction. I currently do not have an explanation for this. LO chain current drops considerably from about 75 mA for the pair of 0885s to 16 mA. I am beginning to really like the MGA 86576!

Dave Mascaro, WA3JUF, recorded his problems with the transmitter section in his club's newsletter, *Cheese Bits*.[4] It includes a detailed description of how to modify the board. I have tried this modification and it works well as described below. A discrete GaAsFET here is a better solution than the GaAs MMIC since much more output power can be obtained. Much of the feedback I have received through the years on both the 2304 and 3456 no-tune transverters has been on how one might get more power out of the transmitter. This is understandable since when these transverters are used *barefoot* as a rover rig and the other station is a fixed home or mountaintop contest station, the limiting factor in making the contact is usually going to be the ability of the fixed station to hear the low power transmitter. Although the noise figure of the original no-tune transverter is a few dB higher than a *performance station*, it is not likely going to be an issue since the fixed station usually runs higher power.

I decided to give an ATF 10135 a try as it looked like a good choice for gain and power output capability. For this device I had to make a few changes to Dave's circuit to bias the FET properly and to ensure stability. The drain resistor was dropped to 5.6 ohms to allow a Vd of 3.6 volts. The source bias resistor needed to achieve 60 mA of drain current is 18 ohms. I had to lower the Q of the gate choke with a series 5.6 ohm chip resistor to stop an oscillation at 11 GHz. With these changes I got an insertion gain of 10.5 dB. When the driver MMICs are replaced with another MGA 86576 as described below, a power output of +17 dBm (50 mW) was obtained. I have not done any extensive testing of stability with different termination SWRs like Dave did on his unit. This issue may best be investigated with a computer modeling program.

No problems have been reported to my knowledge with the two low level driver stages in the transmitter. But having just replaced the balance of the transverter with gallium arsenide, why finish the job with an MGA 86576 in place of the 0185/0285 and then just run the whole transverter off +5 volts? I couldn't resist. The results were excellent for the driver replacement. The MGA 86576 has about 6 dB more gain than the 0185/0285 pair, giving about +6.5 dBm out of the second transmit filter. This higher level is welcome as the desired 3456 MHz signal out of the mixer is reduced slightly due to the lower LO power and the ATF 10135 can achieve greater output with more drive.

Summary

Having done all the above modifications to a 3456 no-tune transverter, I have a stable unit with characteristics listed in the table below. The specs for the original unboxed unit are included for reference.

The total cost of the MMICs for the original unit is about $15. The GaAs replacements will run you only about $28 not counting any onboard or external voltage regulators. A small price for such an increase in performance.

Parameter	Modified*	Original
Noise Figure	2.0 dB	5.0 dB
TX Output	50 mW	8-10 mW
LO Power @ 552	+9 dBm	+16 dBm
Total Current	110 mA	186 mA

Three separate enclosures, absorptive material on cover, MMIC replacements as described in this article.

BIBLIOGRAPHY

[1]Davey, Jim, WA8NLC, "Engineering Considerations for Microwave Equipment Design," *Proceedings of Microwave Update 1993*, ARRL.

[2]Tom Hill, WA3RMX/7, 19335 NW Walker Rd, Beaverton, OR 97006

[3]Wade, Paul, N1BWT, "Improved Battery Regulation for No-Tune Transverters," *Feedpoint*, March/April 1993, North Texas Microwave Society.

[4]Mascaro, Dave, WA3JUF, "3456 MHz No-Tune Transverter Mod," *CheeseBits*, Aug 1991, Pack Rats VHF Society.

Home-Brewing a 10 GHz SSB/CW Transverter

Part 1–Narrowband 10-GHz operation—without exotic or surplus parts—has finally arrived for the microwave builder!

By Zack Lau, W1VT
(From *QST*, May 1993)

Looking for some challenging microwave equipment to build? How about a complete 10-GHz transverter with stability good enough for weak-signal CW work? You don't have to find any exotic pieces to build this project—all of the parts are fairly common. In fact, everything has been available for years.

Despite the transverter's compact package, however, it consists of several modules that you must build. And although the VHF/UHF circuitry follows the no-tune concept developed by Jim Davey, WA8NLC; Rick Campbell, KK7B; and others, the X-band (10-GHz) parts need to be tuned up—preferably with a spectrum analyzer that works through 10.4 GHz.[1]

Design Philosophy

Unlike the no-tune transverters,[2] I decided to develop the transverter as a set of building blocks with stainless-steel or gold-plated SMA connectors. Although this con-

[1] Notes appear at the end of this section.

Table 1

10-GHz Transverter Performance*

Transmit Converter

144-MHz Drive (dBm)	10-GHz Output[†] (dBm)
−10.0	3.8
−3.0	8.5
0.0	10.3
1.0	10.8
3.0	11.6
5.0	12.2
10.0	12.8

Power Output versus Supply Voltage[†]

(Drive signal: −0.8 dBm at 144.06 MHz)

Supply (V)	Output Power (dBm)
10.34	8.5
10.51	10.0
10.75	10.0
12.34	10.0
14.02	10.0

Receive Converter

IF (MHz)	Gain (dB)	Noise Figure (dB)
144	8.82	2.73
146	8.79	2.70
148	8.79	2.71

Noise Figure and Insertion Gain versus Supply Voltage

(IF = 144 MHz)

Supply (V)	Gain (dB)	Noise Figure (dB)
10.3	6.59	3.03
10.7	8.81	2.92
10.9	8.64	2.89
11.3	8.59	2.79
12.4	8.68	2.75
13.4	8.70	2.74
13.5	8.73	2.76
14.9	8.75	2.74

*The data in this table comes from the most recently completed prototype, which consists of the modules described in Part 1 and Part 2 of this article.
[†]Power output was measured with an uncalibrated HP 435B/8481A.

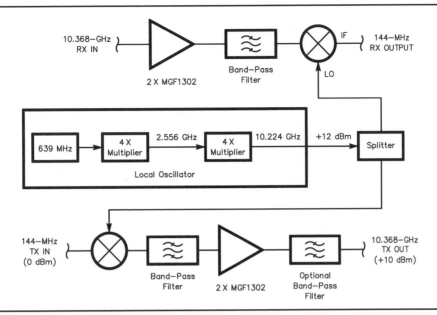

Fig 1—Block diagram of the 10-GHz transverter.

struction method is more expensive and time-consuming than a more integrated approach, it offers several advantages. Most importantly, it allows you to check small portions of the transverter for proper performance. If something doesn't work, troubleshooting is fairly straightforward. And, if you just can't get one of the modules to work, you can simply build another one. Another advantage of this construction method is the shielding that results from packaging circuits in separate boxes. This helps greatly to keep the transverter spectrally clean, with a minimum of spurious outputs and responses. Finally, the transverter is easily updated or expanded to take advantage of improving technology. Making its receiver section state-of-the-art is simply a matter of adding the 1-dB noise figure preamplifier described in December 1992 *QEX*.[3]

A Brief Overview

Fig 1 and Table 1 show the transverter's configuration and measured performance. A local oscillator (LO) feeds a power splitter that drives a pair of mixers. One mixer is used on transmit and the other on receive. The transmit mixer is followed by a filter and amplifiers. A filter following the final stage is optional. Low-noise amplifiers and an image-stripping filter precede the receive mixer. Without adequate image rejection, the receiver sensitivity can degrade by as much as 3 dB.

The Local Oscillator

The most critical part of microwave narrowband work, the LO, starts off with the circuitry developed by WA8NLC and KK7B. The 106.5-MHz oscillator (Fig 2A) is multiplied by six (Fig 2B) to produce a 10-dBm (10-mW) signal at 639 MHz. This signal is then multiplied by four and amplified to 7 dBm at 2.556 GHz (Fig 3).

This is essentially the same scheme used in KK7B's 2.16-GHz LO in July 1989 *QST*,[4] except that I modified the

filters for 639 and 2556 MHz. I also added a 0.47-µF capacitor to provide a low-impedance input for the 78L05 regulator (it can oscillate if not properly bypassed). These circuits are built on fiberglass-epoxy G10 or FR4 PC-board material; the remaining circuits are built on 5880 RT/Duroid.

Choosing the Circuit-Board Material

This part was actually pretty easy: I looked around for something with low enough loss to work well, but that's also readily available to amateurs. The only stock item that meets this description is 0.015-inch-thick (15-mil) 5880 RT/Duroid. This is the same material used in the Tuesday Night Transverter published in the *Proceedings of Microwave Update '88*.[5] The thicker 30-mil 5880 RT/Duroid is definitely unacceptable, as its radiation loss is rather high. Down East Microwave is one possible source.

If availability wasn't an issue, I might have chosen a board thickness that helps to optimize stability via source inductance.[6] Another criteria for choosing board thickness is the interface with the transistors and connectors. Often, it is desirable to minimize the discontinuity between these interfaces by selecting trace widths comparable to the connector diameters and transistor-lead widths. The 15-mil board works pretty well in this area—the 46-mil trace widths fairly closely match the widths of the specified 50-mil chip capacitors.

Crystal Frequency

When choosing an LO crystal, the most important consideration is the crystal's calibration. The tolerance of the International Crystal Manufacturing high-accuracy crystal (#473590) I recommend is 10 parts per million. This means that the crystal can be as much as 1.06 kHz off the marked frequency without deviating from the specified accuracy. Because the LO is multiplied by 96, the transverter's conversion frequency could be as far as *102 kHz* from the ex-

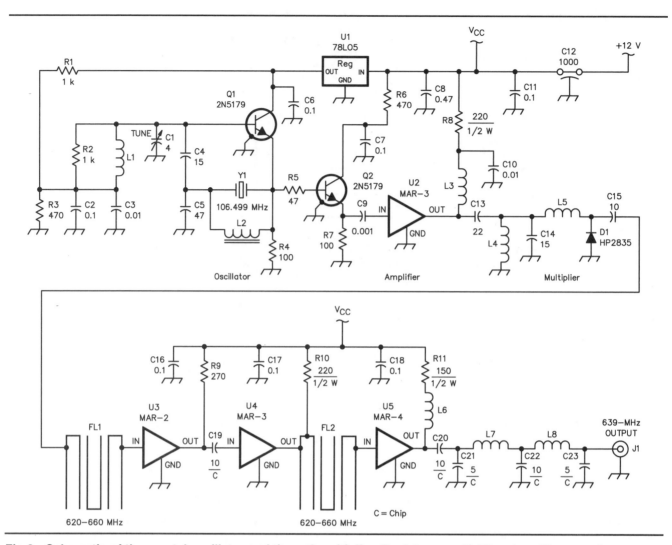

Fig 2—Schematic of the crystal oscillator and times-6 multiplier. Resistors are ¹/₄-W carbon-film or carbon-composition types unless otherwise indicated.

C1—Air-dielectric trimmer capacitor that can be set to approximately 4 pF. Low temperature coefficient is more important than exact value, as L1 can be adjusted to compensate.

C8—Minimum value required to stabilize U1 is 0.33 μF. An electrolytic capacitor can be substituted if proper polarity is observed.

C12—1000-pF feedthrough capacitor. Exact value not critical (100 pF to 0.1 μF should

work well).

D1—Schottky diode. Hewlett-Packard 5082-2835, -2811 and -2800 work well.

FL1, FL2—Band-pass filters printed on PC board.

J1—SMA female chassis-mount connector.

L1, L3, L4, L6—8 turns #28 enameled wire, 0.1-inch ID, closewound.

L2—12 turns #30 enameled wire on T-30-6 toroid core.

L5—5 turns #28 enameled wire,

0.1-inch ID, closewound.

L7, L8—2 turns #28 enameled wire, 0.062 inch ID, turns spaced one wire diameter.

Q1, Q2—2N5179 or BFR91.

U1—78L05 5-V, 100-mA, three-terminal regulator.

U2, U4—MAR-3 or MSA-0385 MMIC.

U3—MAR-2 or MSA-0285 MMIC.

U5—MAR-4 or MSA-0485 MMIC.

Y1—106.499 MHz, fifth-overtone, series-resonant crystal (International Crystal Manufacturing #473590).

pected frequency, even without taking temperature variations into account. Although the oscillator circuit allows some adjustment to compensate for frequency error, attempting to shift the frequency seems to degrade stability.

To make sure that the conversion frequency falls inside the 2-meter band, I specify a 106.499-MHz crystal. Selecting a 106.500-MHz crystal might prove to be unwise if it was cut 10 ppm high—the usual calling frequency of

10.368100 GHz would be just below the 2-meter IF radio's 144.0-MHz band edge—a problem with some radios. You may want to choose another frequency, perhaps even lower, to move the IF to 145 or 146 MHz. If you do this, you'd be wise to investigate possible sources of interference. Keep in mind that hilltops are often pretty bad in terms of interference problems.

The stage following the 639-MHz to 2.556-GHz mul-

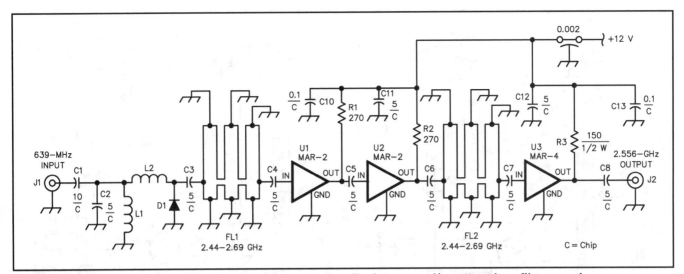

Fig 3—Schematic of the 639-MHz to 2.556-GHz multiplier. Resistors are ¼-watt carbon-film or carbon-composition types unless otherwise indicated.

D1—Schottky diode. Hewlett-Packard 5082-2835 or equivalent.
J1, J2—SMA female chassis-mount connector.
FL1, FL2—Printed band-pass filters.

L1—3 turns #28 enameled wire, 0.062 inch ID, turns spaced one wire diameter.
L2—Printed inductor.
U1, U2—MAR-2 or MSA-0285 MMIC.
U3—MAR-4 or MSA-0485 MMIC.

tiplier is a GaAsFET multiplier, filter and amplifier (Figs 4 and 5) that takes the 2.556-GHz input and provides at least 12 dBm at 10.224 GHz to the LO splitter/mixer board (to be described in Part 2).

Bias Supplies

I know it's not the cheapest way to go, but I decided to build a negative bias supply into each module that requires one (all the stages that use MGF1302s). This reduces the chance of misconnecting the positive and negative supplies. I also opted for active bias supplies, as shown in Fig 5. This figure shows the two equations for calculating components for different bias conditions. For instance, to bias an FET at 3 volts and 30 mA, you first calculate the effect of any resistors used for stability. Often, a 51-Ω resistor is used to stabilize the circuit; if present, it increases the circuit bias voltage to 4.53 volts. One set of standard values that comes close to the bias conditions given above, and accounts for the 51-Ω resistors, is: $R_{dn} = 16\ \Omega$, $R_{an} = 3.6\ k\Omega$, and $R_{bn} = 1.1\ k\Omega$.

I used Intersil ICL7660s to generate the negative bias supplies because they require few external parts. A cheaper alternative is to use NE555 timer chips as oscillators driving rectifiers.[7] I published such a circuit, with a PC-board pattern, in March 1991 *QEX*.[8]

Filter Construction

The transverter's band-pass filters are made from half-inch copper pipe caps, as shown in Figs 6 and 7. These were developed by Roman Wesolowski, DJ6EP; and Kent Britain, WA5VJB.[9] They're affordable, too: You can buy half-inch plumbing caps at home-supply stores for as little as 12 cents each. (Designed to cap pipes that are 0.5 inch ID, these caps actually measure 0.62 inch ID and about ⁹⁄₁₆ inch

long.) I drill and tap the caps (at top center) with #4-40 threads and use nickel-plated brass screws; unplated brass screws should work as well. Kent Britain has forced steel screws through the caps to thread them. Don't use these screws for tuning, though, as steel is unacceptably lossy. I often polish my plumbing caps so that they look nice and solder easily.

A pipe-cap filter ahead of the mixer is adequate in terms of system noise figure, giving an image rejection around 24 dB with a 144-MHz IF. For critical applications, a waveguide filter, such as the one published by Glenn Elmore, N6GN, in July 1987 *QEX*,[10] is recommended. With such a filter, 50 dB of image rejection is easily obtained with a 144-MHz IF. However, for lightweight portable transceivers, plumbing-cap filters seem to be the best compromise. For a clean transmitted signal, you should use one at the final transmit amplifier's output as well.

The completed transverter.

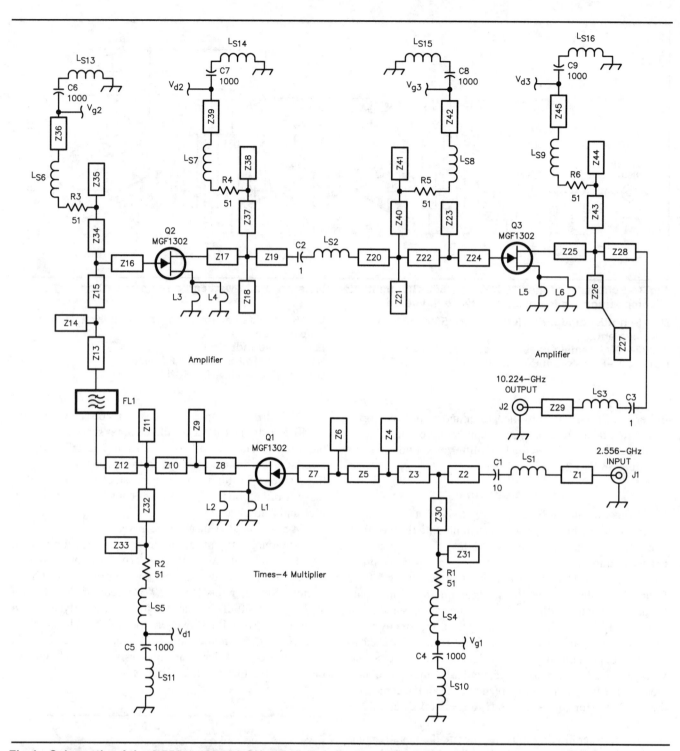

Fig 4—Schematic of the 2.556- to 10.224-GHz multiplier. Resistors and capacitors are chip components. L1-L6 are source-lead inductances. L_{S1}-L_{S16} are stray inductances. Z1-Z45 are etched on the circuit board.

FL1—Pipe-cap filter. See Fig 6. Countersink the ground-plane side of the circuit board hole (by hand) to keep the ⅛-inch UT-141 probe center

conductors from shorting to it.
Q1-Q3—Mitsubishi MGF1302 GaAsFET. Substitution not recommended.

The filters are built on unetched, double-sided, ¹⁄₁₆-inch G10 or FR4 PC-board material. I recommend that you use 0.141-inch semirigid coaxial cable (UT-141) to make the probes. A probe length of about 75 mils is optimum. If you cut them too short—say, 50 mils—the insertion loss climbs from an acceptable 1 to 2 dB to as much as 5 or 8 dB. If the probes are cut too long—say, 100 mils—the image rejection

drops to a measly 10 to 14 dB, though the insertion loss also drops (to 0.5 dB). The probes are spaced ⁵⁄₁₆ inch center to center and the pipe cap is soldered to the ground plane so that the probes are centered within it.

How do you determine the best probe lengths and spacing for pipe cap filters? I developed the filters in this transverter using a spectrum analyzer and trial and error.

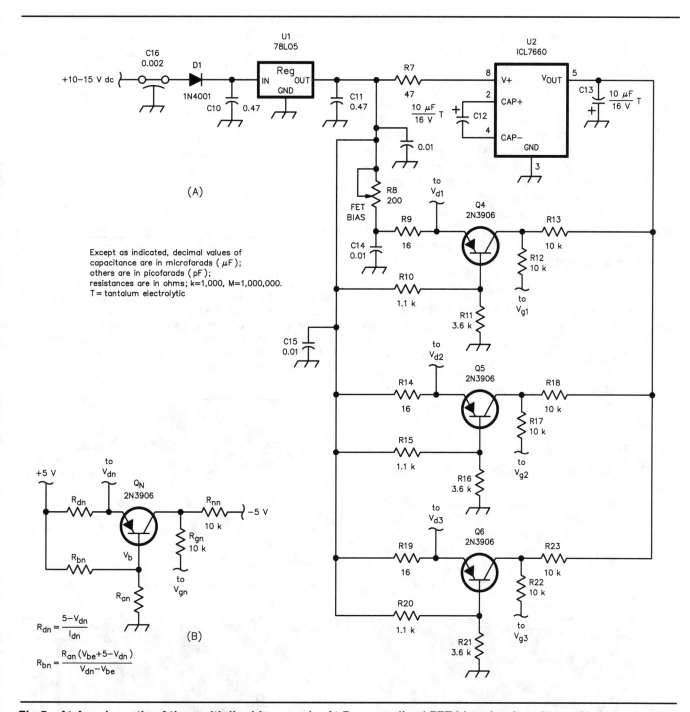

Fig 5—At A, schematic of the multiplier bias supply. At B, generalized FET bias circuit and equations.

C12, C13—Tantalum electrolytics preferred, but
  aluminum electrolytics should work.
C16—Feedthrough capacitor. Value not critical.
D1—Reverse-polarity protection diode.
Q4-Q6—General-purpose PNP transistor; 2N3906 and
  MPS2907 suitable. Plastic-cased devices are easiest

to use.
R8—200 Ω used in the prototypes. 100 or even 50 Ω
  may be suitable.
U1—78L05 5-V regulator.
U2—ICL7660 negative-voltage generator.

The signal source was an X-band mixer and the 10.224-GHz local oscillator. I could have done a lot better with a network analyzer or a scalar sweep setup, but I used what was available to me.

    Filter construction can be fairly critical for optimum performance. In particular, the probes must be accurately cut to length. I estimate my error margin in measuring and

cutting probe lengths to be about 10 mils. The ends of the probes are filed flat, not chamfered or rounded. Filter loss seems to be a few tenths of a decibel lower with the dielectric left on the probes, although it is easier to trim the probes exactly with the dielectric removed.

    You may be tempted to use 0.085-inch semirigid cable because it's easier to handle than UT-141. A similar filter I made using this material gives 24 dB of image rejection, but has 3.4 dB of loss. The probe length for this cable is 70 mils. A filter using 100-mil probes of 0.085-inch semirigid cable has only 2 dB of loss, but the image rejection drops to a barely acceptable 17 dB. UT-141 is better for this application.

I recommend that you assemble the cable and solder it to the ground plane *before* measuring and cutting the probe length. Otherwise, the length may change as you work on the cable. With these filters, a potential problem is caused by the center conductor moving around slightly, particularly when the cable is straight and the center conductor forms the center contact at the connector end. Bending the cable helps to prevent this problem, but the best solution is to use connectors that captivate the center conductor, keeping it from being pushed inward.

Enclosures

As shown in the photo on page 2-27, I use 0.025-inch-thick, half-inch-wide brass sheet stock to make the enclosure walls. Instead of soldering SMA connectors to the walls, I attach them with #2-56 screws; either method is acceptable. The 25-mil brass stock is ideal for tapping small screw holes. Other commonly available thicknesses can also be used, although 20-mil stock is a bit flimsy and 32-mil stock is more difficult to solder.

Duplicating the Circuit Boards

Using PC-board layout software, I've developed artwork for each of the transverter's circuit boards. To make it as easy as possible for *QST* readers to build this transverter, ARRL HQ is making the circuit-board artwork available in three forms: as PostScript files downloadable from the ARRL HQ telephone BBS; as negative film for those with access to photographic methods of circuit-board production; and as laser-printed positive images that can be trans-

ferred directly to the PC-board material.[11]

Several methods are available for transferring toner from the laser engine to the circuit board. Plain paper is my favorite.[12] Start with a clean circuit board (roughed with 400-grit sandpaper) and a laser-printed reversed positive

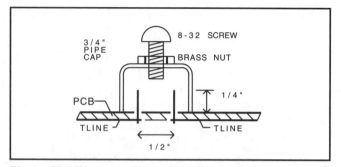

Fig 6—FL1 Pipe-cap filter. The probes are made out of the center conductor and dielectric of UT-141 coax—use the dielectric to help hold the wires in place—the length from the ground plane to the tip is 125 mils. The ground plane is carefully countersunk to avoid shorting to the probes.

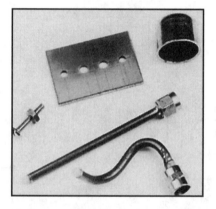

Fig 7—The pieces that make up a 10-GHz band-pass filter, before assembly. (photos by Kirk Kleinschmidt, NTØZ)

Fig 8—At left, a photo of the 106.5-MHz oscillator circuit, built ground-plane style on the back of the 639-MHz multiplier board. Oscillators built this way exhibit stability an order of magnitude better than etched PC-board versions. This is especially important for minimizing drift at the 96th harmonic of the oscillator frequency—10.224 GHz. At right, a top-side view of the 3³/₄- × 4³/₄-inch assembly shows the 639-MHz etched band-pass filters.

image of the board on plain paper. Then use an ordinary household iron at its linen setting to iron the image onto the board. Buffer the iron from the sheet of paper with the pattern on it with a second, clean sheet of paper. Run the iron over the board in a pattern that uniformly heats the material for 30 seconds or so for the 15-mil Teflon boards and at least a minute for the G10/FR4 boards. The iron's heat liquefies the plasticized toner and fuses it to the circuit board.

After ironing, place the board and paper (now fused to the board) into plain water for a few minutes, then remove it from the water and carefully rub away as much of the paper as you can. If the transfer process leaves incomplete traces, clean the board again with sandpaper and start over with a new copy of the artwork. You can correct minor imperfections with an etch-resist pen and carefully cut pieces of Scotch tape. Cover the bottom (ground-plane) side of each board with Scotch tape, then etch the boards. Peel off the tape and remove the toner with plain steel wool.

Oscillator Construction

I didn't develop circuit-board artwork for the 106.5-MHz local oscillator. If you want stability adequate for a 10-GHz SSB/CW system, a quartz-crystal-controlled system is marginal—you really can't throw away any stability to make construction easier. Remember: *The LO is multiplied by 96 before being mixed with the 144-MHz IF signal!*

You could use a double-sided circuit-board layout, except that stability is 10 times worse than that of a ground-plane version. So, I opted for the ground-plane version (Fig 8A). I also used a high-stability, air-dielectric trimmer at C1, as some ceramic trimmers have a high temperature coefficient. The trimmer value isn't critical, as L1 can be adjusted to compensate.

I recommend that you build and align the oscillator as follows. Build the oscillator with a 47-Ω, ¼-watt resistor in place of the inductor/crystal combination (L2 and Y1). When you power up the circuit, tune C1 so that the oscillator operates at 106.5 MHz. After replacing the 47-Ω resistor with the crystal and its resonating inductor, verify that the oscillator starts reliably as power is applied. A minor adjustment of C1 may be necessary for reliable starting. I don't recommend trying to adjust C1 for a given oscillation frequency.

The 639-MHz to 2.556-GHz multiplier (Fig 9) has no tuning adjustments. You simply verify that its power output is between 5 and 10 dBm.

Amplifier Design

I chose to use MGF1302 GaAsFETs for all the 10-GHz circuits. These seem to be the most readily available, low-cost parts that work well at this frequency. The transverter uses seven of them, and they cost less than $7 each from several

Fig 9—The 1¼ × 5-inch 639-MHz to 2.556-GHz multiplier board uses MMICs to provide a 5-mW filtered intermediate LO signal that drives the 10.224-GHz multiplier.

Fig 10—At left, the business end of the 2⅛- × 2⅜-inch 2.556- to 10.224-GHz multiplier module. At its far left, the UT-141 probes couple to the pipe-cap filter shown at right. The FET biasing circuits are also shown at right.

sources. Ideally, a transverter like this would use 10-GHz MMICs for gain blocks, but these weren't available during project development. Not only were the available packaged GaAs MMICs too pricey (around $40 each), but they weren't designed to work at 10 GHz. The second choice was the Avantek ATF13735, but commercial purchasers have made the standard part the short-leaded ATF13736, which is more difficult to use than the long-leaded version. I prefer to use devices with long leads since they're easier to install in circuits that use lead inductance as a circuit component.

Multiplier and Amplifier Construction

In each of the transverter's building blocks, I build the RF circuitry on one side of the ground plane and the biasing circuitry on the other. After etching the boards, I drill and countersink holes for the power leads. Also countersink the ground-plane foil around the multiplier board's filter-cable holes so that the UT-141 center conductor doesn't short to the ground plane. Countersink the holes *by hand* with a relatively large drill ($\frac{3}{16}$ to $\frac{1}{4}$ inch). It's important to do this manually—you'll be surprised how easy it is to drill through such thin, soft material! Cut slots for the FET source leads as discussed in the next section. Then add the brass walls and install the connectors. Build the biasing circuitry after the transistors are installed.

Circuit performance may be improved slightly, as discussed in the next section, though the design is relatively broadband and should operate adequately despite minor construction variations. Computer simulations predict gain flatness within a decibel across the 10-GHz band.

I glue RF-absorptive rubber or foam to the insides of the enclosure lids. This reduces the chance of waveguide effects disrupting circuit operation.[13]

GaAsFET Installation Tips

Beware of soldering irons with significant ac leakage. People blow up lots of devices because their soldering iron tips aren't at ground potential. Measure your soldering iron's tip-to-ground potential if you have any doubts.

The circuits in this transverter use the GaAsFET source-lead inductance as a circuit component. Use the photos as guides when installing them. Bend the source leads down at the ceramic device body, then insert them into holes carefully cut in the circuit boards using a #1 X-ACTO blade or similar weapon, as is done in *The ARRL Handbook*'s GaAsFET preamplifiers.[14] Be sure to cut the holes so that the device is centered on the board traces. Once the device is installed, bend the source leads up flush with the bottom of the board and solder them to it.

Of course, take the usual precautions when handling GaAsFETs, which are static-sensitive. Chapter 24 of *The ARRL Handbook* discusses these practices.

Adjusting the 2.556- to 10.224-GHz Multiplier

First, adjust the filter-tuning screw for maximum output. Next, set the bias trimmer for maximum power output. You may then want to tune the amplifiers. Do this using a tuning tool made out of a $\frac{1}{16}$-inch-square piece of thin copper sheet or

foil stuck into the end of a piece of Teflon tubing. Slide the tool along the input and output lines, looking for hot spots—places where the presence of the foil makes the power output increase. After finding them, turn off the power. Next, solder a piece of foil at each hot spot and adjust its position with high-quality tweezers.

Coming in Part 2

When you finish building the blocks described here, you'll have a clean 10.224-GHz local oscillator. Next, I'll describe the mixer/splitter board and the preamplifier/power amplifier circuit, and some 10-GHz antenna ideas.

Notes

[1]It *may* be possible to tune up the system using a Gunnplexer unit with an S meter, but I haven't attempted it and can't guarantee that it will work. A Gunnplexer should be able to pick up a properly functioning LO even with lots of attenuation between them.

[2]*QST* has published a series of no-tune transverters developed by KK7B and WA8NLC, including versions for 903, 2304, 3456 and 5760 MHz. The most recent of these is J. Davey, "A No-Tune Transverter for the 2304-MHz Band," *QST*, Dec 1992, pp 33-39. See the notes at the end of D. Mascaro, "A High-Performance UHF and Microwave System Primer," *QST*, May 1991, pp 30-33, for details on the others.

[3]Z. Lau, "The Quest for 1 dB NF on 10 GHz," RF, *QEX*, Dec 1992, pp 16-17.

[4]R. Campbell, "A Clean, Low-Cost Microwave Local Oscillator," *QST*, Jul 1989, pp 15-21.

[5]K. Bailey, R. Larkin and G. Oliver, "TNT for 10 GHz," *Proceedings of Microwave Update '88* (Newington: ARRL, 1988), pp 80-95.

[6]Amplifier stability is affected by the inductance of the FET source leads.

[7]I've also used little surplus boards with surface-mount NE555s, although some of these are poorly constructed and had to be resoldered for reliable operation.

[8]Z. Lau, "Power Supply for GaAsFET Amplifier," *QEX*, Mar 1991, pp 10-11.

[9]K. Britain, "Cheap Microwave Filters," *Proceedings of Microwave Update '88,* pp 158-162.

[10]G. Elmore, "A Simple and Effective Filter for the 10-GHz Band," *QEX*, Jul 1987, pp 3-5, 15.

[11]The ARRL BBS can be reached at 860-594-0306 (1200/ 2400, N, 8, 1); one 250-kbyte file, KH6CP10G.ZIP, contains all the PC-board artwork. Send paper and film artwork requests to the Technical Department Secretary, ARRL, 225 Main St, Newington, CT 06111. Request the MAY 1993 *QST* KH6CP 10-GHz TEMPLATE and be sure to indicate whether you need paper or film artwork. The template package also includes part-placement diagrams for the transverter's circuit boards. There is a charge of $2 for ARRL members and $4 for nonmembers. Please enclose a check made out to ARRL.

[12]J. Grebenkemper,"Ironing Out Your Own Printed Circuit Board," *QST*, July 1993, pp 42-44.

[13]K. Britain, "Works Great! Until You put it in the Box?" *Proceedings of the 25th Conference of the Central States VHF Society* (Newington: ARRL, 1991), pp 33-34. See Note 5 for ordering information.

[14]L. Wolfgang, ed, *The ARRL Handbook for Radio Amateurs,* 1993 ed (Newington: ARRL, 1992), pp 32-22 through 32-38.

Home-Brewing a 10 GHz
SSB/CW Transverter

Part 2—Designed to work with last month's 10.224-GHz local oscillator, this month's mixer, power amplifier and pre-amplifier round out your narrowband 10-GHz transverter.

By Zack Lau, W1VT
(From *QST*, June 1993)

In Part 1, I described a 10.224-GHz local oscillator (LO) designed to drive a dual-mixer board like those used in the no-tune transverters. If you've completed the modules described in the first section, you should have a working 10.224-GHz LO. The mixer board described this month contains a two-way etched power splitter that delivers equal LO signals to the transmit and receive mixers, which are also etched on the same PC board. On transmit, one of the mixers combines a 144-MHz IF signal with the LO to generate a 10.368-GHz signal; on receive, the other mixer combines the incoming 10.368-GHz signal with the LO to produce a 144-MHz IF output. An external pipe-cap filter (described last month) in each 10-GHz mixer line eliminates the image, passing only the desired signal. Two-stage GaAsFET amplifiers of the same RF design, but using different bias settings, serve as a 10-GHz preamplifier and power amplifier.

The first section also shows the transverter block diagram, and covers construction techniques and etching-pattern availability for the transverter's circuit boards, component sources, and performance data for the finished transverter.

Mixer Construction and Tweaking

If you're building transverters from surplus hardware, the most difficult module to obtain is not the LO, but the mixer. Builders have gotten widely varying results, even when copying the same design. For most people, 10 GHz is just too high a frequency to accurately build a no-tune mixer that works well. The difficulty is that a full-wavelength microstrip transmission line is only 0.6 inch long at 10 GHz. So, a typical rat-race mixer (which requires signals to be 180° out of phase for proper cancellation) really needs to built with tolerances under 0.005 inch (5 mils).

This problem has several solutions. One is to simply accept the inferior performance. Usually, the conversion loss isn't too bad if you copy a known-good layout, but the LO rejection relative to the PEP output signal can be as little as 10 dB. For receive purposes, LO rejection really doesn't

make much difference.

A better solution is to tune the mixer. Once you've etched and assembled the mixer board, terminate all ports in 50-Ω loads or sources. You don't want to look at the mixer through an image-reject filter, unless it is properly tuned. Otherwise, the mixer and filter tuning will interact, making it difficult to adjust the mixer for proper operation. I normally connect the mixer to the LO, attach a 0- to –10-dBm, 50-Ω 144-MHz source at the IF port, and a spectrum analyzer at the RF port.

Fig 11—The splitter/mixer board. The IF ports (at top) are connected to the mixers via 0.01-μF encapsulated chip capacitors. Ceramic capacitors can be used instead. The mixers use Hewlett-Packard HSMS 8202 Ku-band diode pairs. The HP HSMS 2822 2-GHz diode pair can be substituted, but this device has more conversion loss at 10 GHz than the HSMS 8202. Mixer tuning for best LO rejection is done by adding a small piece of foil at point A or point B. *(photos by Kirk Kleinschmidt, NTØZ)*

I usually adjust the LO rejection first. This is done by placing a small piece of copper foil at point A or B indicated in Fig 11—at either side of the junction between the 70-Ω ring and the LO-input line. This shortens the transmission line slightly on one side. Usually, the LO suppression improves with the copper at one point and worsens with it at the other point. True, the copper foil mismatches the amplitude slightly, but this is better than having an improper phase shift. Usually, LO rejection is 17 or 18 dB below the saturated output (this equates to the specification-sheet figure of 27 or 28 dB of LO-to-RF port isolation). Keep in mind that even a lid covered with absorptive rubber or foam affects the tuning slightly. You don't want to tune the mixer to perfection only to have to retune it after installing a cover.

I find that the obtainable LO rejection depends on how well I made the board. Mixer rings that look almost perfect often allow 5 or 10 dB better rejection; ones that look as if they were drawn quickly with a crayon may be almost impossible to tune (though they often work just fine for receive).

Finally, tune the mixer's RF port for maximum output into a 50-Ω load (as described in Part 1 under "Adjusting the 2.556- to 10.224-GHz Multiplier"). I've been unable to etch mixers consistently, so all of my mixers are a little different.

Three short wires, 0.21 inch of #28 enameled wire, serve as 10-GHz RF chokes and 144-MHz shunts at the mixer board's RF and LO inputs (Fig 11).[15] This improves the isolation between the mixer's IF ports. Without them, there is little to stop a 2-meter signal from crossing the power divider. Adding these wires increases the isolation between the IF ports from an almost negligible 4 dB to a decent 40 dB.

It shouldn't be necessary to tune the load termination, though you may want to. As you might guess from the layout, I tacked on the radial stub to ground the 51-Ω chip resistor. Purists may want to use a 68-Ω resistor and tune out the reactance to get a really good 50-Ω load at 10 GHz (as is done in the TNT). If you have them, you can also use 50-Ω microstrip terminations for this; I've gotten them from surplus isolators.

[15] Notes appear at the end of this section.

10-GHz Power Measurement

Measuring RF power at 10 GHz presents a challenge; calibrated measurement devices can be very expensive. Fortunately, measuring *relative* power requires only a diode detector and a sensitive dc voltmeter. The 10-GHz power measurements required to optimize this transverter needn't be absolute; relative power measurement is acceptable. A convenient way to measure power in a 50-Ω system is to couple some RF into a low-offset Schottky diode, such as a Hewlett-Packard 5082-2835 (commonly used as a microwave frequency multiplier), filter its dc output, and measure this voltage with a high-impedance voltmeter. This measurement approach gives useful output down to the milliwatt level. Of course, you can also use a commercial diode detector rated to 10 GHz.

To build a detector, etch or cut a 50-Ω microstripline on a small piece of Rogers 5880 RT/duroid with 1-oz copper cladding (the same material used in the transverter's 10-GHz circuits). See Fig A. A 50-Ω trace is 46 mils wide (0.046 inch) on this material. Terminate the microstrip in SMA connectors and enclose the board with brass strip for rigidity, like the transverter. Mount the diode and other components as shown in Fig A. Fig B shows the equivalent circuit. The length of the diode lead that runs along the

50-Ω stripline affects the amount of RF energy coupled into the diode, as does its spacing from the microstrip trace.

This detector can be used for tuning the transverter's multiplier, filters and amplifiers. To use the detector, terminate one end in a 50-Ω load that's good to 10 GHz. (Alternatively, you can substitute a 50-Ω microstrip load for one of the SMA connectors.) Couple RF into the other port via a 3- to 10-dB attenuator, to ensure that the circuit under test is terminated with a stable 50-Ω load. Measure the voltage on the feed-through capacitor using a sensitive voltmeter or oscilloscope.—*Kent Britain, WA5VJB*

Fig A—This 2- × 1¼-inch diode detector gives useful dc output for 10-GHz power measurement down to about 1 mW. It uses a low-offset Schottky diode (such as the HP 5082-2835), with its anode lead soldered to the ground plane. Its cathode lead follows the 50-Ω microstrip trace for about ¼ inch and is spaced about ⅛ inch from the trace (neither dimension is critical; a longer lead and closer spacing increase coupling). A 1- to 10-kΩ resistor, also soldered to the cathode lead, routes rectified energy to a feedthrough capacitor.

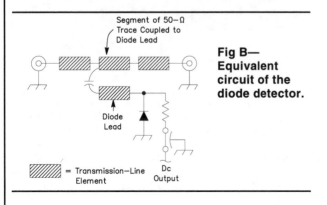

Fig B— Equivalent circuit of the diode detector.

Segment of 50-Ω Trace Coupled to Diode Lead

Diode Lead

Dc Output

▨ = Transmission-Line Element

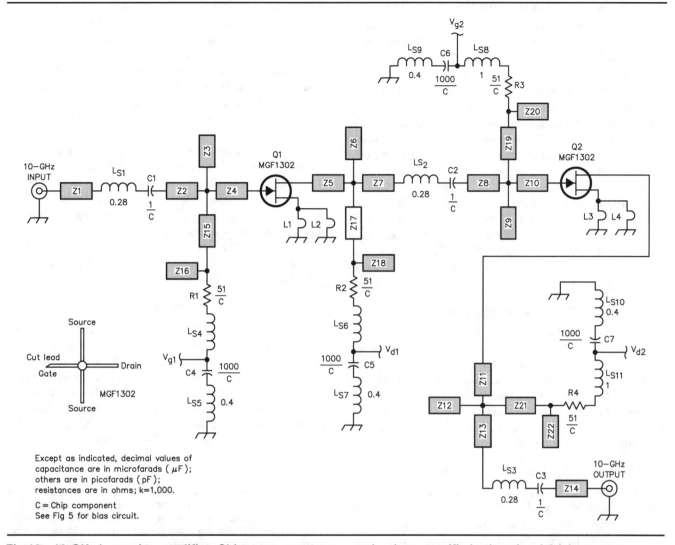

Fig 12—10-GHz low-noise amplifier. Chip components are used unless specified otherwise. L1-L4 are source lead inductances. L_{S1}-L_{S11} are stray inductances (in nanohenries). Z1-Z22 are etched on the circuit board.

C1-C3—1 pF. Use high-quality, 50-mil ceramic chip capacitors such as ATC 100As.

Q1, Q2—MGF1302. Substitution not recommended. Set bias at 10 mA and V_{ds} = 3.0 V for low-noise receive preamplifier operation. For the transmit amplifier, set bias at 30 mA and V_{ds} = 3 V. For additional biasing information, see the text and Fig 5.

Fig 13—Top and bottom-side views of the completed 10-GHz two-stage amplifier. Microwave-absorptive foam is glued into the lid to suppress waveguide propagation modes inside the amplifier enclosure, which could provide enough feedback from output to input to cause oscillation. The amplifier's bias circuitry, like that of the 10.224-GHz multiplier (Figs 4, 5 and 10) is built on the bottom side of the amplifier board for convenience. External biasing is also acceptable and may be more convenient.

Power Amplifier and Preamplifier

The same RF design is used for the transmit and receive amplifiers (see Figs 12 and 13). Tripling the bias current from the 10 mA used in the receive-side amplifier to 30 mA in the transmit amplifier increases the circuit's 1-dB compression point from 5.7 to 10.8 dBm. Gain increases from 18 to 19 dB and the noise figure rises by about 1 dB. Computer-modeling results indicate that the Rollett stability factor, K, drops a little, but since it's still above 3 (a K greater than one denotes a stable design), this shouldn't be a problem—even if the amplifier is terminated at the input and output with a high mismatch (such as sharp filters).

It may be possible to get a bit more output by increasing V_{ds} to slightly more than the 3 volts I used, but this would require redesigning the bias circuit (Fig 5). Like many transistor amplifiers, this amplifier's saturated output, typically 14 dBm, is more than twice the recommended output for linear operation.

System Integration

To complete the transverter, build two band-pass filters as shown in the first section. You can tune them with the aid of the diode detector described in the sidebar, but a few minutes with a spectrum analyzer makes the process easier.[16] Then, following the block diagram of Fig 1, assemble the transverter's blocks. Connect a suitable IF radio, check to make sure the transmit converter and receive converter operate, and you're on the air!

Antenna Thoughts

Most people looking for a high-gain antenna end up with some sort of parabolic reflector. If you put a low-gain horn antenna in the right spot in front of a reflector that is anything close to a parabola, it will probably outperform anything of similar dimensions on this band. People have used everything from metal snow sleds to trash-can covers, in addition to more obvious choices such as light collectors and surplus military/commercial gear. Commercial sources for new dishes exist, but even small dishes are expensive when purchased new. Look for a surplus reflector.

Perhaps the simplest antenna I've seen is a quarter-wave monopole—with a piece of sheet metal as the ground

plane! The most complicated is undoubtedly a loop Yagi—it works, but it is more of a curiosity than a practical way of getting 18 to 20 dBi of gain. A horn is much easier to make. *The ARRL Antenna Book*, *The ARRL UHF/Microwave Experimenter's Manual*, the RSGB *Microwave Handbook, Volume 3* and various VHF/UHF/microwave conference proceedings contain duplicable designs. Chapter 18 of the RSGB *Microwave Handbook, Volume 3*, contains all the information you need to get started.

I have yet to adjust one of my 10-GHz antennas with an SWR meter, yet I've made lots of 10-GHz contacts of more than 200 km. Usually, if I do any tweaking at all, such as adjusting the location of a dish feed, it has been for maximum received signal. Similarly, I've adjusted my coax-to-waveguide transitions this way, adjusting tuning screws for minimum loss. Of course, even SWR is no indication of how well an antenna really works. The *real* test is to compare antennas and see which does best.

Summary

Although it takes some effort to build, the transverter described in this two-part article provides useful and exciting 10-GHz SSB and CW capability. Perhaps the best part is that you don't have to hunt through flea markets to find a surplus "brick" LO and filters, or deal with any of the other traditional hassles of getting on this fun band! What hilltop will *you* operate from in this year's ARRL 10-GHz Cumulative Contest in August and September?

Notes

[15] Part-placement diagrams with component values and more detail for each of the transverter's modules are part of the template package obtainable from the ARRL Technical Department Secretary. See Note 11, page 2-32 for details.

[16] When tuning these filters, you can use the finished 10.224-GHz LO and a power meter (see the sidebar) to make sure that the filters aren't tuned to the LO or image frequency. To tune a filter, first connect it to the LO and adjust the tuning screw for maximum output at the LO frequency. Then adjust it for peak response at 10.368 GHz by connecting it to the transmit-amplifier output and backing the tuning screw *farther out of the filter cavity* while looking for maximum filter response.—*Ed.*

An Image-Phasing Transverter for 10.368 GHz

By Doug McGarrett, WA2SAY

(From *The 22nd Eastern VHF/UHF Conference*)

When all the local 10-X group started talking about using a 2-meter SSB input to their 10 GHz transverters, *and* using a separate 2-meter sideband radio for liaison, it started to sound expensive. In addition, when you put your power amplifier on a standard mixer, you waste half of your capability amplifying your image, unless you build a filter at the RF output frequency and insert it before the power amplifier. Why not use brains, instead of brawn, I thought.

The image-phasing mixer is a technology which goes back to the early days of single sideband radio.[1] Back when the Central Electronics 10-B was the way most people got on SSB, that was the technology they used. So there's nothing new here. Just a bit higher in frequency, that's all.

The magic of the image phasing mixer is that, as an up-converter, you need no filter to get rid of your image frequency. And the usual double balanced mixers suppress the local oscillator pretty well also. As a down-converter, you again need no preselector to keep out the image frequency noise. Typical image rejection is 30 dB or better for narrowband operation.

The basic device is shown in Figure 1, along with an HF transceiver, a local oscillator source, and an antenna, as well as incidental equipment. This is not going to net you any tremendous DX, but it has made a 2-way QSO over a 24 mile obstructed path! Think what it will do when it has RF amplifiers going in both directions between it and the antenna! The RF amplifiers will be a receiving preamp with about 1.5 dB noise figure and 15 dB or so gain, and a transmitting amplifier with better than 1 watt output and about 25 dB gain. Add microwave relays, a dish antenna, and stir.

Meanwhile, the mixer: In order to achieve image phasing, you need 90° phase shifters at one of the RF ports (not both) and the IF port. The other RF port must be fed either in phase or 180° out of phase. Fortunately, the devices to do this are fairly commonly available on the surplus market. (Except for the IF hybrid, which can be home-brewed.) My IPM is made up of two WJ (Watkins-Johnson) "MinPac"

M80D double balanced mixers, fed with an Omni-Spectra 2032-4096-00 quadrature hybrid, and a Narda 4315-2 Wilkinson hybrid. All of this is hooked together as directly as possible. One of the hybrids has male connectors on one side, the other is connected using male-male SMA adapters.

I discovered (by Murphy's Law) that you have to have the quadrature hybrid on the antenna side. Otherwise, you generate upper sideband, but receive lower sideband. (Note that what I mean by "sideband" in this application is the frequency relationship between the local oscillator and the signal at the antenna port.) I have not done the vector analysis to determine why this happens, but I can testify that it does! (Later research indicates that the quadrature hybrid is always shown connected at the antenna end of the mixer. He who does not study history is doomed to repeat it, they say.)

The IF hybrid was made using a one-to-one transformer wound on a small ferrite toroid, with capacitor coupling across the input to output connections at top and bottom (Figure 2). If this is done right, you wind up with a 3-dB hybrid directional coupler with the classic 90° phase shift between the two output ports. This device can be found in the ham literature, as well as other places.[2]

The performance of the mixer was evaluated for use directly on an antenna, since that is how I knew I would first have it hooked up. With a local oscillator drive signal of +19 dBm, and an IF signal of +10 dBm, I could get +4 dBm out of the mixer with reasonably good linearity. If I didn't care so much about linearity, I could push the RF up to about +15 dBm, and get about +7 or +8 dBm out. That's fine for CW. The image rejection (at +10 dBm IF drive) was 34 dB. The LO rejection was 17 dB. Noise figure with a mediocre IF preamp was 11 dB.

The above was written as a quick report; since then, Bruce Wood, N2LIV, has asked me to expand on it somewhat for the Proceedings of the Eastern VHF/UHF Conference. One of the items not discussed in detail is the IF hybrid. As shown by the reference, the first article to discuss this, as far as I'm aware, was by Reed Fisher,

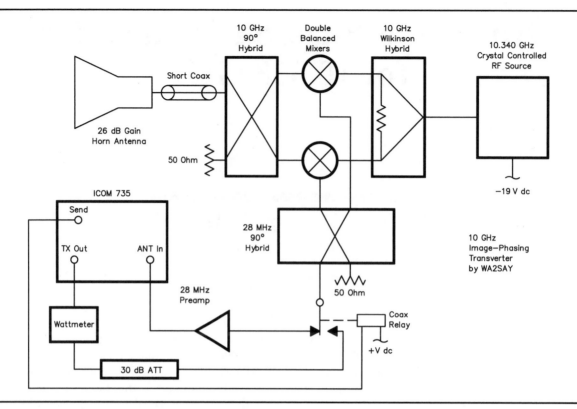

Figure 1

W2CQH. He followed the IRE paper with an article which appeared in *Ham Radio*, I believe, but my copy was loaned out to someone who needed it more than I did. (That someone also got my copies of Norgaard's and Dome's papers on single sideband and phasing networks.)

The trick is to wind a very tightly coupled twisted pair of wires on a small toroid form, in a fashion similar to that of Ruthroff's transformer.[3] In this case, the twisted pair transmission line which results will be terminated in a capacitor at each end. The design is obtained by means of these equations:

$$Z_0^2 = L/C$$

where Z_0 is the impedance at each port of the coupler, L is the inductance of one winding on the core, and C is twice the value of the terminating capacitors. (See Figure 2.)

The frequency at which the device operates with 3 dB coupling is determined by setting

$$\omega L = Z_0 \text{ or, } L = Z_0/2\pi f$$

and the terminating capacitance, C/2, is determined by

$$C/2 = 1/(4\pi f Z_0)$$

The toroidal core should be a material of relatively low permeability, such as FaiRite type 67 material. I optimized the transformer design using EEsof Libra for Windows, and then tweaked using a network analyzer. In the end, the phase shift was just shy of 90°, so I cheated and added about an inch of RG188/U cable to one port.

It is possible to buy the if hybrid, if you choose the right frequency. Both Toko and Mini-Circuits sell 90° hybrids ready made. I don't know what the cost of the Toko version is, since I can't find a source who will sell in "onesies."

Pennstock (a distributor) has them, but they also have a minimum order of $100. Mini-Circuits also has 90° hybrids. The price is pretty steep, but they will sell single units to individuals who live outside New York State. (They can't handle the required in-state sales tax.) The Toko part number for a 28.35 MHz center-frequency hybrid is B4QF-1004. The Mini-Circuits part number for a 23 to 40 MHz hybrid is PSCQ-2-40. If you choose this route, all you have to do is mount the hybrid on a little piece of circuit board and add SMA connectors.

The rejection performance of the image phasing mixer is critically dependent on the equality of the power division and the accuracy of the 90° phase shift. The equation that defines the performance is

$$\text{Image rejection} = (1 + 2\cos\phi + R^2) / (1 - 2\cos\phi + R^2)$$

where ϕ is the phase error from quadrature, and R is the amplitude ratio, in volts.

A simple computer program[4] (in Microsoft or GW

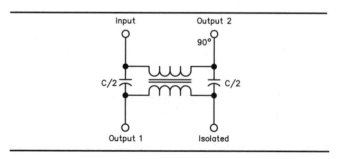

Figure 2

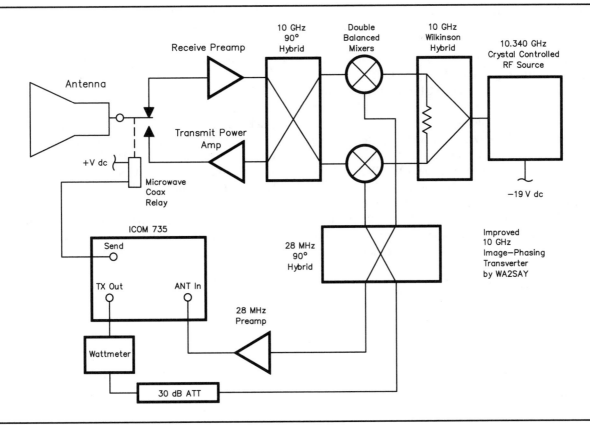

Figure 3

Transverters 2-39

Appendix: IPMREJ.BAS

```
100 REM IPMREJ FINDS IMAGE REJECTION
    IN AN IPM
110 REM ACCORDING TO K2RIW :
120 CLS
130 PRINT" IMAGE REJECTION PROGRAM FOR
    AN IMAGE PHASING MIXER"
140 PRINT
150 PI=3.14159265#
160 INPUT" DB DIFFERENCE";D
170 INPUT" PHASE DIFF, DEGREES";DEG
180 REM CONVERT DB TO VOLTAGE RATIO
190 K=10 ^ (-D/20)
200 REM CONVERT DEGREES TO RADIANS SO
    BASIC CAN DO COS ON IT
210 THETA=DEG* PI/180
220 PARTIAL=(2*K/(1+K^2))*COS(THETA)
230 REJ=(1+PARTIAL)/(1-PARTIAL)
240 DBREJ=10*LOG(REJ)/LOG(10)
250 DBREJ=INT(DBREJ*10+.5)/10
260 PRINT
270 PRINT" IMAGE REJECTION IS
    " DBREJ" DB"
280 PRINT
290 END
```

Basic) is provided in an appendix, which will allow analysis of the effect that any combination of amplitude and phase error will have on the performance. (The mixers, of course, must be carefully constructed to meet the necessary criteria. Fortunately, most good microwave balanced mixers designed for your frequency of interest will work well, with a minimum contribution to system error. The insertion loss, however, is directly dependent on the mixer insertion loss characteristics.)

The above discussion does not mention the need for RF gain in both the transmit and receive directions. While it is a simple matter to configure single-pole double-throw RF relays at each end of the mixer, and at the junction of the antenna feed with the RF preamp and the RF power amp a little inspection of the circuit leads me to the conclusion that you can build the whole transverter with only a single coax relay! Look at Figure 3. If you took only one hybrid and used the "other" port, you would get the wrong sideband.

But if you use the "other" port on both hybrids, you should get the same sideband. Since these ports are otherwise isolated, the scheme should work, but I confess that I have not yet tried it.

Afterword

The ICOM 735 radio and the 30 dB pad have been replaced by a modified Radio Shack Realistic all-mode HTX-100 10 Meter radio. RF is taken before the final, which has been disconnected, thru a 13 dB attenuator. The receive IF signal is fed into a new, direct input port to the receiver in the radio.

The form suggested in Figure 3 has been implemented. The RF preamplifier is a modified TVRO LNA-down-converter module, using just the 10 GHz front end portion, for a noise figure of <2 dB and a gain of about 34 dB. A short waveguide section was added at the preamp input, and tapped for 3 brass 0-80 screws, which were adjusted to bring the noise figure down from 3.5 dB.

The RF power amplifier is now an 8-W TWT, satellite surplus. This unit has about 40 dB gain, and a circulator and small pad is used at its input. Note that the RF quadrature hybrid must be terminated in good impedance matches at all ports if this scheme is to work with proper image rejection. The same is true of the IF hybrid; the 13 dB of padding on the 28 MHz transmitter output port is applied right at the hybrid.

Finally, the antenna is now a 30-inch diameter dish, fed by a $3/4$-inch circular waterpipe waveguide "shepherd's crook" feed with a Chapparel launcher.

References:

[1] Norgaard, D. E., "The phase-shift method of single-sideband signal generation," and "The phase-shift method of single-sideband signal reception," Single Sideband Issue, *Proc. IRE*, vol. 44, Dec. 1956.

[2] Fisher, R. E., "Broad-Band Twisted-Wire Quadrature Hybrids," *IEEE Trans. on Microwave Theory and Techniques*, May 1973.

[3] Ruthroff, C. L., "Some Broadband Transformers," *Proc. IRE*, vol 47, Aug. 1959.

[4] Knadle, R., K2RIW, private communication.

BUILDING BLOCKS AND CONSTRUCTION

A Simple T/R Sequencer

Protect sensitive RF circuits from burn out by sequentially switching from transmit to receive.

By Zack Lau, W1VT

(From *QEX*, October 1996)

What, another T/R sequencer? Yes, I decided to make it as simple as possible, while adding +12-V outputs and reverse polarity protection. I think the reverse polarity protection is pretty nifty—who hasn't worried about hooking up batteries backwards and frying something? The 12-V outputs make it real easy to wire up my transverter designs, since they are made up of modules that run off 12 V. Hopefully, this design will spur more designers into adding reverse polarity protection.

The primary reason for using a sequencer is to protect the RF relay and the amplifiers hooked up to it. RF relays can be damaged by hot switching at high power levels. Unfortunately, what constitutes high power is rather fuzzy—I have not found any good references that adequately address this topic. Based on my experience with microwave transverters, I always put in a sequencer when switching more than half a watt. I don't bother with them in inexpensive systems running less than 100 mW.

Like the T/R sequencer designed by Chip Angle, N6CA, this sequencer uses a quad comparator to monitor the voltage on a charging or discharging capacitor.[1] I looked at using an integrator to get more uniform delay intervals, but decided to stick with the simpler circuit. An integrator can generate a nice triangular waveform, as opposed to the exponential curve generated by an R-C network. The 10-Ω resistors are used to provide hysteresis, so the outputs don't switch back and forth unnecessarily near the transition point. They are supposed to provide a little positive feedback, instead of the more common negative feedback used in other applications.

This sequencer first turns off the receiver, then activates the relay, amplifiers and finally the transmit IF outputs when switching from receive to transmit. When going back to receive, it turns off the transmit IF, amplifiers, and relay before reactivating the receiver. The idea is to introduce enough delay between these states to allow everything to settle down, and to reverse the order when switching from transmit to receive. This is especially important with a mechanical relay which may make intermittent contacts for a few ms when switched.

A big advantage to using solid-state amplifiers is that you can switch them off during receive. Not only does this reduce the possibility of hot switching, it reduces the chance of amplified broadband noise getting into the receiver. Using a PIN diode switch to cut off transmit drive also helps to prevent hot switching—I use the transmit IF signal to control this switch.

Since I've never needed to change the polarity of the output of one of my sequencers, I decided some simplification was in order. Instead of the XOR gate Chip used, I decided to use a hard-wired switch based upon the principles of a transfer switch. To switch polarity with this design, you change a pair of resistors from horizontal to vertical, or vice versa. By labeling the blank area of the board between the resistors, identifying the resistors ought to be straightforward. The reduction in parts count ought to enhance reliability.

I've also taken advantage of the improvements in switching transistor technology. The International Rectifier P-channel IRF 9Z34 will easily switch 2 A, enough to power a 5-W GaAs FET power amplifier. Similarly, the Zetex ZTX 789 in a little TO-92 style case will actually switch a small SMA relay that draws a few hundred mA. The bonus to using more expensive PMOS/PNP parts is that the switched supplies are reverse polarity protected. I just needed to protect the comparators with a diode and use bipolar electrolytic capacitors which don't care about voltage polarity. Cheap 2N3906s are used for the RX and TX IF supplies, since they typically draw under 100 mA. You could use ZTX 789s instead of the 2N3906s, with the appropriate bias resistors, for higher current.

You may want to replace Q2 with a VN10LP N-channel FET. This will allow you to hook up the PTT line without pulling the voltage down significantly. Another

[1]Notes appear at the end of this section.

advantage is that you can now hook up the PTT input to a stiff voltage source without frying Q2. But these FETs aren't as easy to get as common NPN switching transistors.

Of course, it makes sense to watch out for even newer FETs, which will undoubtedly offer better performance at lower cost. There is even a trend toward lower gate thresholds, which allow better performance at lower voltages. For instance, the IRF 7104 drops only 154 mV when sourcing 0.57 A ($V_{gs} = -5$ V). But, despite the marketing hype,

you can do even better than "full enhancement." It only drops 117 mV with a V_{gs} of -10 V.

Construction

I made the pads big enough to accommodate swaged terminals, which are a really nice way of making dc connections if you have the tooling to rivet them to a fiberglass circuit board. The board is a bit crowded—I wanted the board to fit nicely on the wall of a chassis box only two

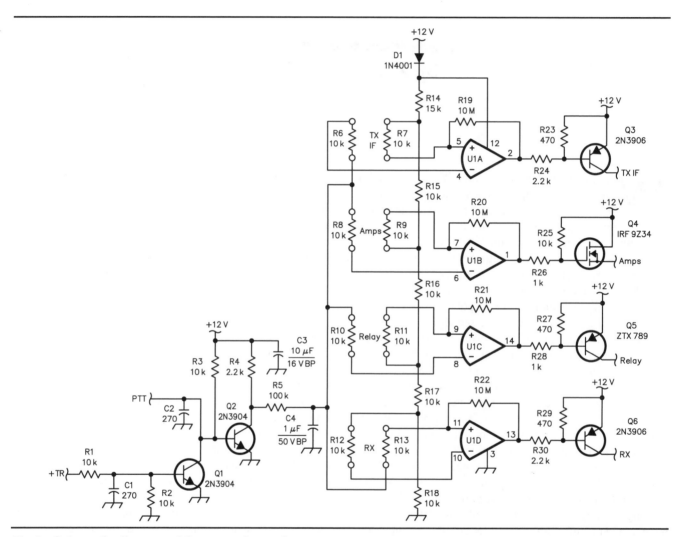

Fig 1—Schematic diagram of the transmit-receive sequencer.

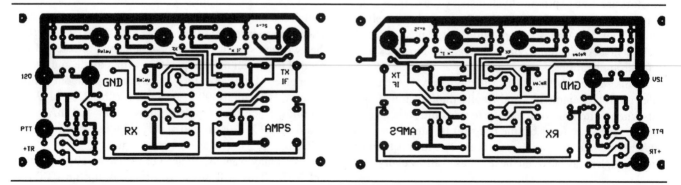

Fig 2—Etching pattern for the transmit-receive sequencer circuit board.

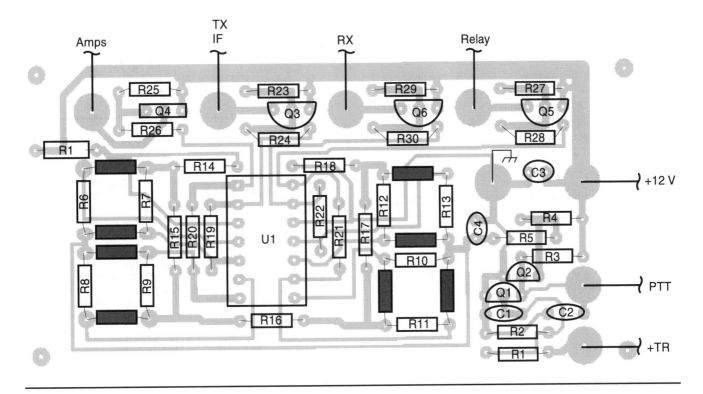

Fig 3—Parts placement diagram for the transmit-receive sequencer circuit. The shaded resistors indicate alternate positions for R6 to R13 to invert the signal sense. See text.

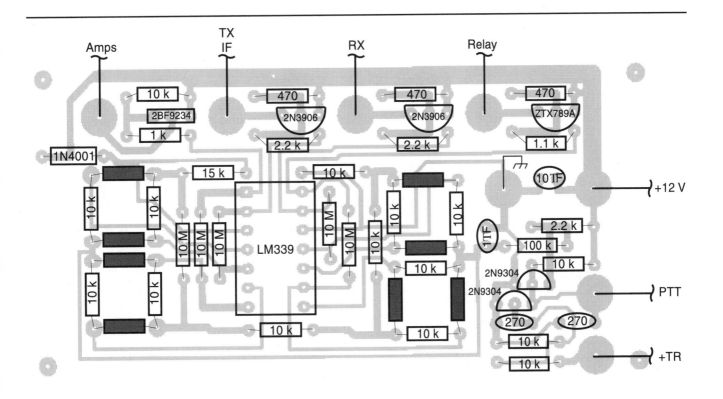

Fig 4—Parts placement diagram using component values and part numbers.

inches high. A mirror image of the etching pattern is provided—it simplifies the toner transfer process that some people use to make circuit boards. Similarly, a parts placement diagram using part values and component designations is also provided in Fig 4.

To test the board, I made a little fixture out of LEDs and dropping resistors. A separate fixture makes it easy to line up the LEDs in the proper sequence. It may be useful to slow down the sequencing by bridging the 1-μF timing capacitor with a 10-μF capacitor. This makes it easier to see the LEDs turn on and off.

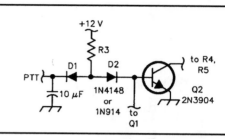

Notes

[1]Angle, Chip, N6CA, "TR Time-Delay Generator," *The ARRL Handbook for Radio Amateurs*, 1997, pp 22.53-22.56.

Feedback

Unfortunately, the T/R sequencer doesn't actually protect its outputs from reversed polarity, though the sequencer itself survives reverse polarity. Incandescent test lamps instead of LEDs are recommended for anyone experimenting with such circuits. The problem isn't easy to solve—power FETs contain a parasitic diode across the junction. Normally, protective circuits get around this problem by reversing the source and drain connections, but this doesn't work here. That fix would disable the use of the FET as an on/off switch—it would be effectively on all the time.

Mario Miletic, S56A, suggests the improved PTT input circuit above. He also suggests that the decoupling capacitors C1 and C2 be increased to 10 nF, though the author thinks that 270 pF is more appropriate for a VHF/UHF/SHF station.

A "Fool-Resistant" Sequenced Controller and IF Switch for Microwave Transverters

Feeling foolish since you blew up that nice new transverter because the T/R switching wasn't sequenced right? Here's a way to avoid that problem.

By Paul Wade, N1BWT
(From *QEX*, May 1996)

Last summer, I suffered the failures of two 10-GHz preamps and one coax relay. Fortunately, none occurred at critical times, and I finally rigged up an inconvenient but safer two-switch scheme to prevent further problems. But I did resolve to come up with a better control system this winter. Ideally, it would be fool-proof, but fools are too resourceful for that, so I've tried to make it as fool-resistant as possible.

Discussion

For several years, I've been using variations of a transverter IF switch by KH6CP.[1] This has worked well in several of my transverters, and I've made various improvements, but it does not adequately sequence various switching functions.

Three sequencing techniques are commonly used. The first is to intercept the PTT line so the transceiver is controlled by the switch box. Often, this requires modification of the transceiver, particularly those that do break-in CW only, transmitting when the key is touched. I want to be able to interchange various transceivers without modification so I can lend spare equipment to willing rovers.

The second approach uses a fixed-sequence switch, usually a series of time delays, which, once started, go through the sequential operations without further safeguards.

The last, and least successful, method is to switch an external relay directly from a transceiver's PTT line. Often, the current available from the PTT line is inadequate for driving a relay, and I know of several cases where the transceiver has been damaged by this technique.

My preference would be a switch that goes through sequential operations but checks that appropriate conditions are met before proceeding to the next step—in logic design, this is called a *state machine*. As I started to sketch out the sequence of operations I wanted, I realized that ordinary T/R switches, such as relays, don't have an appropriate state to deal with break-in transceivers, which are delivering RF before the switch is ready for it. To deal with this, I use a

PIN-diode IF switch and have designed the controller to have a third state, in addition to transmit and receive, in which all applied RF power is absorbed. I call this third state the *safe* state. Since one of the functions of the IF switch is to attenuate the transmitted power from the IF transceiver, the safe state is implemented by adding two PIN diodes to absorb the power.

Design

The first step in the design process is to sketch out the desired timing for the switching sequence. This evolved to the timing diagram shown in Fig 1, which goes through one cycle from receive to transmit and back to receive. The second step is to synthesize a logical state machine that generates the desired timing. The final step is to actually design a circuit that implements the logical state machine. Following this progression helps to ensure that the final circuit will operate as intended since there is a clear target to work toward.

The desired operation of the state machine sequence is shown in the *state diagram*, Fig 2. The system starts in the inactive receive state. When an activation signal is received, the system moves to the *safe* state, absorbing all RF power. The switching sequence can then continue at whatever speed is required, not releasing the RF power to the transmit circuitry until the system is ready to go to the transmit state. Normally, this would mean removing power from the receive section, then driving the microwave T/R relay, waiting long enough for it to switch (or to sense that the fail-safe contacts closed, if you are fortunate enough to have a relay with this feature) and finally, applying power to the transmit section. When the activation signal is removed, we go from the transmit state to the safe state, reverse the switching procedure, then return to the receive state.

Since this state machine is intended to be used in several transverters and with various IF transceivers, I added some options to increase flexibility:

1. RF sensing ensures that any RF power applied to the IF port will cause switching to the safe state—even if no control signal is applied—to protect the transverter from

---

[1] Notes appear at the end of this section.

damage. Full RF switching may be enabled by setting the J17 jumper, allowing the use of any transceiver, even a hand-held, for the IF.

2. PTT polarity selection is provided since some transceivers ground the PTT output on transmit while others provide a positive voltage. There are separate inputs for these two PTT polarities; each requires low current and has a switching threshold of about 5 V.

3. Single-cable switching supports transceivers that put the dc PTT voltage on the RF output cable. Jumper J3 sets the PTT polarity for the IF cable.

4. A transmit-ready signal can be sensed. Some amplifiers require a warm-up period, so this input must be grounded to indicate that everything is ready to transmit; otherwise, the switching sequence will remain in the safe state and not continue. This input could be automatically or manually switched.

5. Fail-safe sensing detects when the fail-safe contacts on a coax or waveguide relay have closed—and prevents transmitting until they do. I've not yet found a good coax relay with this feature, but my 10-GHz waveguide relay does have it. Jumper J15 selects between fail-safe operation and time-delay-only sequencing.

6. FET output drivers for the safe state activate a coax relay, activate dc power switching and drive LED indicators for the operator. I like to have three LEDs: TRANSMIT READY, SAFE, and TRANSMIT.

7. A DPDT relay may be jumpered to be switched by any of the FET drivers to operate at the desired point in the desired sequence.

The components for each of these options are indicated on the schematic diagram, Fig 3, and may be populated as desired.

Circuit Description

Some have suggested using a small microprocessor to implement switch sequencing. The flexibility and programmability of this approach would be great, but I am very cautious about putting microprocessors in high-intensity RF fields. Since my intent is to include both the PIN-diode switch and the controlling state machine in a small metal box, there may be a significant amount of RF in the box. Therefore, I chose to design using components that are cheap, proven and readily available, and are also slow enough not to respond to RF. And wherever possible, I use these components in circuits I have used before and know to work well.

Let's take a quick tour of the schematic diagram, Fig 3. The IF transceiver connects to J1, and its transmit power is reduced by the attenuator, R1, R2 and R3. The values shown provide about 14 dB of attenuation. Since low-inductance power resistors are becoming hard to locate, it may be necessary to adjust the attenuator design to fit the available component values. I described how to do this using a computer program, PAD.EXE, in *QEX*.[2] The program, which calculates resistor values and power ratings for attenuators, is available from the *QEX* Web site, **http://www.arrl.org/files/qex/** in file qexpad.zip. The input attenuator used here is designed for the 2 to 3-W output available from small portable transceivers.

The attenuator is followed by the PIN-diode switch. A PIN diode acts as an RF conductor when dc is flowing through it, but acts as an RF open circuit when reverse-biased. Each PIN diode in this circuit is supplied with +6 V at one end, so the other end may be switched between +12 V and ground to reverse the bias. D1 and D2 select the transmit or receive path on the IF side, while D5 and D6 select the path on the transverter side. The transmit path goes through an adjustable attenuator that can be adjusted for 20 to 38 dB of total transmit attenuation. This is needed because most mixers require around 1 mW or less of power. The receive side uses an MMIC amplifier stage, A1, to overcome the loss of the input attenuator—the MAR6 provides enough gain to end up with 6 dB of net gain ahead of the transceiver, with a noise figure better than that of most transceivers.

The safe state is provided by PIN diode D4, which shorts the output end of the attenuator. Turning off FET Q4 causes current to flow through D4, making it an RF conductor, and causes D5 to be reverse-biased, making it an open circuit for RF. Thus, RF flowing into the transmit path has no output path and must be dissipated in the attenuators. The reflected power must pass through the attenuator twice, for a total loss of 60 dB, so essentially no reflected power is seen by the IF transceiver.

An additional safety feature is provided by D3, which is turned on by FET Q2. D3 shorts out any transmit energy that leaks through D2 (when it's off) and also disables MMIC amplifier A1 by reducing the dc voltage supplied to it.

The switching states for the PIN-diode switch are straightforward: in the receive state, FET Q3 is turned on, which

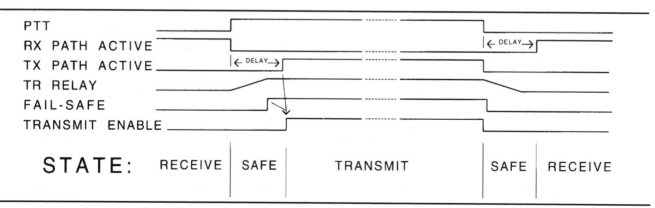

Fig 1—Sequencer timing diagram.

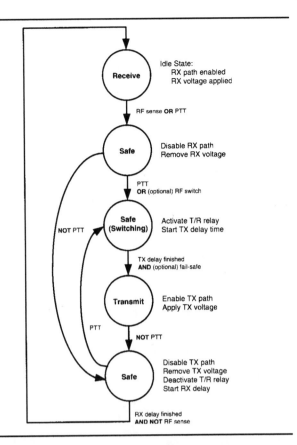

Fig 2—Sequencer state diagram. The circles show the individual states, while the text to the right of each circle shows the actions performed upon entering the state. The text next to each connecting line shows the conditions necessary to advance to the state the line goes to.

causes D6 to turn on. FETs Q1, Q2 and Q4 are turned off, so D2 and D4 are on while D1, D3 and D5 are off—only the receive path is active. The next state is the safe state, reached by turning on FETs Q1 and Q2 and turning off Q3; this turns on D1 and D3 while turning off D2 and D6, so the IF input is switched to the transmit side but the output side is not connected and D4 shorts the attenuator output. Finally, the transmit state is reached by turning on Q4, which turns off D4 and turns on D5, completing the transmit path.

With the resistor values shown, the PIN-diode currents are adequate for an input power level of about ¹/₈ W, so the input attenuator should reduce the IF power to this level or lower. For higher powers, it would be necessary to increase the on current through the diodes, particularly D1. However, it shouldn't be necessary to switch much power, since the RF output to the mixer input should be milliwatts or less.

All the FETs used in this circuit are N-channel enhancement-mode MOSFETs used as switches. The sources are all grounded and the gate is the control element. When the gate voltage is close to the source voltage, or ground, the FET is turned off, and no current flows from drain to source. To turn the FET on, the gate voltage must be several volts more positive than the source voltage, allowing current to flow from drain to source with only a few ohms of resistance. For practical purposes, we may consider the drain to be shorted to ground when the FET is on and open circuited when the FET is off. Since these are insulated-gate FETs, no gate current is possible and no dc power is required for switching.

The gate voltage required to switch the smaller FETs is roughly 2.5 to 3 volts, but larger power FETs such as Q14 require a somewhat higher voltage, so the whole switching circuit operates at 8 V, provided by the three-terminal regulator, IC3.

The rest of the schematic describes the control logic. The RF-detect section, from C13 to Q5, drives IC1A to force the system to the safe state as soon as RF input is detected. The PTT section, from Q6 and D9 to Q7, is a DTL (diode-transistor logic) gate with a switching threshold set by Zener diode D10. The PTT output is inverted by IC2A to also drive IC1A and force the safe state. Note that IC1A is drawn as an OR gate, with inversion bubbles on the inputs to show that they are asserted low; thus the output of IC1A is asserted when either input is in the low, or asserted, state. The output of IC1A is inverted by IC2B which activates OR gate IC1B, thus driving the *RX Disable* signal to turn FETs Q1 and Q2 on and put the PIN-diode switch in the safe state. The output of IC1B is also inverted by IC2C to drive the *RX Enable* signal. This turns FET Q3 off when Q1 and Q2 are turned on, and vice-versa. Finally, IC1B also drives FET Q10, which is turned on in the safe and transmit states so it may be used as a signal to control the voltage supplied to receive stages and preamps.

The PTT section has two inputs, PTT-L on J4 and PTT-H on J5. PTT-L must be grounded, or asserted low, to activate, while PTT-H requires a positive voltage, or high assertion. Both inputs have an operating threshold in the 2 to 5-V range, so any input voltage below the threshold is considered low and any input above the threshold is considered high. The high threshold provides considerable tolerance for different rigs, dirty contacts, etc. The PTT section can also be activated through the IF cable input on J1—any dc voltage on J1 is delivered to the logic circuit through RFC2. Jumper J3 selects the polarity for the IF input; the right-hand position selects PTT-L and the left-hand position selects PTT-H.

The transmit-ready section, from J6 to Q8, is another DTL gate. Its output drives IC1C, which is drawn as a NAND gate; both inputs must be asserted high for the output to be asserted low. The other input to IC1C is selected by jumper J17; in the lower position, it is the output from the PTT circuit. Thus the IC1C logic function requires both transmit ready and PTT to be asserted. The upper jumper position takes the output from IC1A, which also includes the RF detection, making the logic require both transmit ready and either PTT or RF detect. This allows switching using only RF detection. Capacitor C15 sets the hang time for RF switching. With the values shown, switching time seems fast for SSB or for slow CW, so a bit of experimentation might be needed to find a time that feels right.

The output of IC1C is inverted by IC2E (note the inversion bubble on the input, to match the output of IC1C which is asserted low) to drive FETs Q13 and Q14, one of which should be used to enable the T/R relay. The IC1C output also drives FET Q11, which is an inverter with a time delay set by R28 and C18. Q11 drives the *TX Enable* signal, so completion of the time delay turns on FET Q4 to allow the transmit power to flow through J2. When jumper J15 is in the upper position, the completion of the time delay will also allow IC1D to switch, driving Q12 and enabling the transmit state. The lower position of the jumper forces IC1D to wait until J14 is grounded by the fail-safe contacts on the T/R relay.

When PTT is released and no RF is detected, the output of IC1A is deasserted. This voltage transition passes through FET Q9, an inverter with a time delay set by R24 and C16. Until the time delay completes, pin 13 of OR gate IC1B remains asserted, keeping the PIN-diode switch in the safe state while all the other switches are released. Since the safe state prevents any RF from getting through, sequencing of the switches isn't critical in this direction.

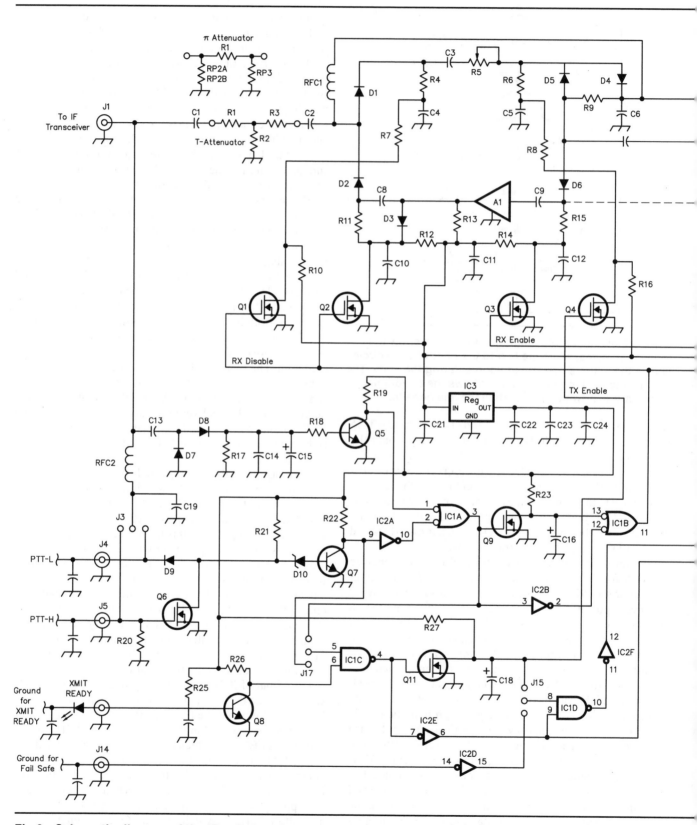

Fig 3—Schematic diagram of the IF switch and sequencer. (See Table 2 for parts list.)

Finally, IC3 regulates the logic voltage to 8 V to maintain constant time delays. The R and C values specified yield time delays of 200 to 300 milliseconds, but the delay can be increased or decreased by changing the values. For instance, increasing C16 from 10 μF to 16 μF would increase the time delay by about 60%. Alternatively, increasing R24 from 33 kΩ to 51 kΩ would have the same effect.

Relay RLY1 may be driven by Q13 to operate at the same time as the T/R relay, during the safe state, or driven by Q12 to operate when entering the transmit state. Notice diode D12 across the relay coil. This serves to protect the FET from the reverse voltage spike caused by removing the current from the relay coil. All relay coils should have a diode to protect the driving circuitry; even a relay driving another relay can suffer contact damage from the switching spike.

Construction

I decided that this circuit is complex enough to justify layout of a printed-circuit board since my intent is to use copies in several transceivers. All the components between the two vertical rows of jacks on the schematic diagram, Fig 3, fit on the PC board. A double-sided board with plated-through holes was needed for full interconnection; the top-layer pattern is shown in Fig 4, and the bottom layer is in Fig 5. Boards are available from Down East Microwave.[3]

All the chip capacitors are mounted on the bottom of the board. I chose to put the PIN diodes on the bottom also to keep lead lengths short in the RF path. All of the other components are on the top side of the board, as shown in Figs 6 and 7. Note that the smaller power FETs, such as Q1, have inconsistent pin-outs that vary with part number and manufacturer. Check the data sheet, and make sure that the source lead connects to ground, which is the wide trace running all over the top of the board. The gate lead connects to the middle pad of each footprint, leaving the drain at the far end.

Component values are not critical. I've tried to calculate optimum values, but any resistor or capacitor value could be changed to the next higher or lower standard value without significant effect. The RF diodes are stocked by Down East Microwave. All other components are readily available from Digi-Key.[4] The cost of all components totals less than $15, not counting the enclosure box and connectors.

The PC board is sized to fit inside a small die-cast aluminum box since a shielding enclosure is highly desirable.

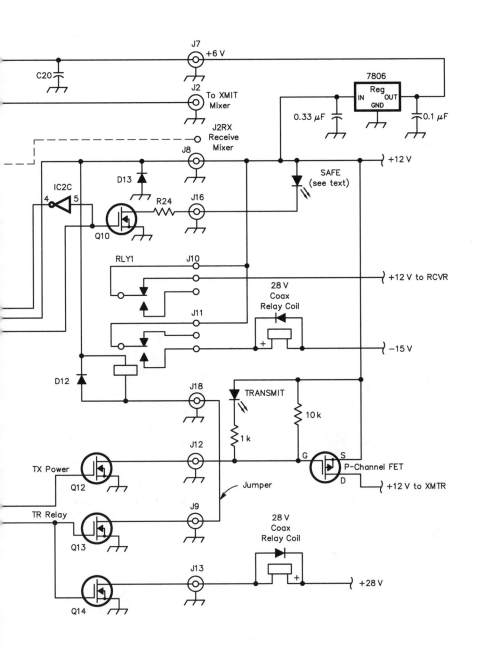

Application

There is enough flexibility in this circuit that using it requires some decisions; on the other hand, it should be possible to fit it to your system needs rather than forcing the system design to match the controller. The portions of the schematic diagram outside the two vertical rows of jacks show some of the possible functions.

The first decision is whether the transverter uses a single mixer, as shown in Fig 8, or separate mixers for transmit and receive, as shown in Fig 9. A single mixer would connect to J2; otherwise, the transmit mixer connects to J2 and the receive mixer connects to J2RX, which is a hole in the PC board next to C9. In this case, D6 and R15 must be removed, and Q3, C12 and R14 may be removed or used for another switching function, as described below.

The next decision involves the control signals. I usually provide inputs for both polarities of PTT using different connector styles (RCA phono for PTT-H, subminiature phone for PTT-L). The transmit-ready and fail-safe inputs can go to connectors if they are used. Otherwise, they should be jumpered to ground to avoid floating inputs. Finally, jumpers J3, J15 and J17 must be installed as described in the circuit description. The switch will not operate without these jumpers.

Finally, we must decide how to use the control outputs. I chose to only provide outputs grounded by FET switches, except for the floating relay contacts, to keep unwanted voltages off the board. The signals that drive the FET switches are labeled on the schematic to indicate function. Some possibilities that I have used are shown in the right-hand side of the schematic. The internal relay, RLY1, can be driven by jumpering J18 either to J9, timed to switch a coax relay, or to J12, timed at the transmit state of the sequence. An external coax or waveguide relay usually requires 28 V for operation, which can be provided from a +28-V supply and switched with the larger power FET Q14, or connected between +12 V and a –15-V supply and switched with the internal relay contacts since many transverters already generate a negative voltage internally.

Power for the transmit stages may be switched with the internal relay contacts or with a solid-state switch using a P-channel power FET like the IRF-9130 or IRF-9530, which can switch several amperes with a small voltage drop. Note that the P-channel FET is used "upside-down," with the positive voltage connected to the source, as shown in the schematic, since a P-channel FET operates using voltages opposite those of the N-channel FETs described above.

Receive stages and preamps may be switched in several ways. The schematic shows the simplest, using the internal relay to disconnect the voltage at the same time that the T/R relay operates. A more robust sequence would be to remove power when entering the safe state; FET Q10 would be the appropriate driver, with R24 replaced with a jumper so the connection is directly to J16. FET Q10 is turned off during receive and on in all other states. If the receive voltage is set by a variable three-terminal voltage regulator, connecting the adjust pin of the regulator to Q10 would turn off the regulator output. Another alternative, for transverters with separate mixers for transmit and receive, would be to use FET Q3, which turns on during receive and off in all other states, the inverse of Q10. In the two-mixer configuration described above, Q3 is not needed, and R14 and R15 are removed so their pads are available as connection points.

Of course, one needn't be constrained by the printed wiring. If one of the FET switches is not used for the function shown, it can be used for a different function by connecting its gate to the appropriate switching line. All it takes is a hobby knife to cut the trace and a soldering iron to add a wire.

LED indicators may be driven by any output and can be driven by the same FET that drives a relay since the additional current is small. The schematic shows a TRANSMIT READY LED in series with J6, so grounding the transmit-ready line draws enough current to light the LED. If there is no LED in this line, R25 could be much larger to reduce current drain.

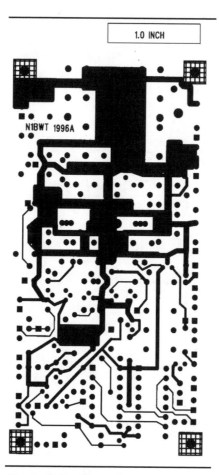

Fig 4—Top-layer printed-circuit board pattern.

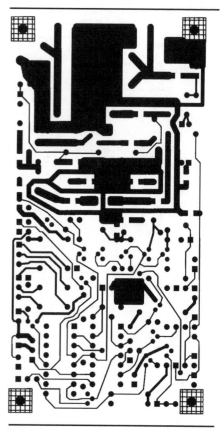

Fig 5—Bottom-layer printed-circuit board pattern. (Note: Board requires plated-through holes.)

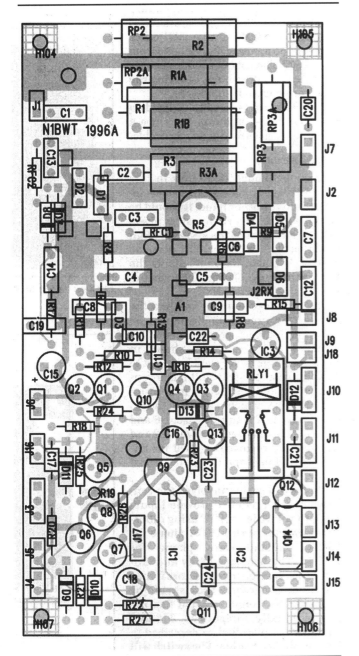

Fig 6—Parts-placement diagram for the printed-circuit board.

Fig 7—The top side of the PC board. The small power FETs are Motorola BS170.

The PIN-diode switch requires +6 V to operate, which may be obtained from a three-terminal regulator if not otherwise available. This regulator easily fits inside the die-cast box, as can be seen in Figs 8 and 9.

Finally, all lines entering and leaving the box should be properly filtered. I strongly recommend a bypass capacitor on the inside of the box and a ferrite bead on the wire between the capacitor and the PC board, plus a ground wire from box to board for each connection. I've seen equipment lacking these components unable to operate properly in the high-intensity RF environments found at many mountaintop sites. Listening to TV sync buzz all day is no fun!

The RF connections at J1 and J2 must have closely coupled grounds from box to board; twisted-pair or coax is preferred. The mounting standoffs do not provide an adequate ground path for RF.

Performance

I have built five of these switches and made RF measure-

Fig 8—A completed unit, built for use with a single-mixer transverter. The small power FETs are Siliconix VN2222.

Fig 9—A completed unit, built for use with separate transmit and receive mixers. The small power FETs are Zetex BS170.

Table 1
Measured Performance

Frequency	Receive Gain	Safe-Mode Gain	Transmit Gain
30 MHz	5.5 dB	−60 dB	−31 dB
50 MHz	5.5 dB	−63 dB	−31 dB
144 MHz	6.0 dB	−59 dB	−30 dB
222 MHz	6.5 dB	−50 dB	−29 dB
432 MHz	7.0 dB	−35 dB	−26 dB

ments on three of them over the range of frequencies normally used for transverter IFs, with the trimpot set for 30 dB of transmit attenuation at 144 MHz. The results shown in Table 1 are typical.

Clearly, the PIN-diode switch works well at up to 222 MHz, with more than 50 dB of attenuation in the safe state and about 6 dB of gain in the receive state. The trimpot range for setting total transmit attenuation was from 20 to 38 dB at 144 MHz. The switch is still usable at 432 MHz as long as the voltage supply to the transmit amplifiers is sequenced to augment the reduced attenuation in the safe state.

The RF-detect circuit operates reliably with the output from an IC202 transceiver, roughly 2 to 3 W, switching smoothly and ignoring glitches like double-clicking the mike button by

remaining in the safe state. I added attenuation between the IC202 and the switch to reduce power. The RF-detect circuit continued to operate with 15 dB of attenuation, at a power level of about 100 mW, but not with 21 dB of attenuation, or roughly 25 mW. This should be adequate margin for safe operation.

Conclusion

The IF switch described here is sequenced to provide fool-resistant operation and is flexible enough for most transverter applications. This combination should make microwave operation more reliable and successful and help protect the environment by reducing the amount of smoke being released from our equipment.

Notes

[1]Lau, Z., KH6CP, "A VHF/UHF/Microwave Transverter IF Switch," *QEX*, August 1988, pp 3-4.

[2]Wade, P., N1BWT, "Building VHF Power Attenuators," *QEX*. April 1994, pp 28-29.

[3]Down East Microwave, 954 Rte 519, Frenchtown, NJ 08825, tel: 908-996-3584, fax: 908-946-3072.

[4]Digi-Key Corporation, 701 Brooks Ave S, PO Box 677, Thief River Falls, MN 56701-0677, tel: 800-344-4539 (800-DIGI-KEY), fax: 218-681-3380. Also on the World Wide Web at **http://www.digikey.com/**.

Table 2
Parts List

A1	MAR-6 MMIC
C1,2,3,4,5,6,7,8,9,10, 11,12,14,15, 19	470 to 2000-pF chip capacitor
C13	2.2-pF disc capacitor (at 50-432 MHz)
C15	22-μF electrolytic capacitor
C16,18	10-μF electrolytic capacitor
C17,20,22,23,24	0.1-μF capacitor
C21	0.33-μF capacitor
D1,2,3,4,5,6	1SS103 PIN diode
D7,D8	1N5711, 1N5712 or HP5082-2035 hot-carrier diode
D9	1N914 or 1N4148 small-signal diode
D10,11	5.1 or 5.6-V Zener diode (1N751, 1N752, 1N5231 or 1N5232)
D12,13	1N4001 rectifier diode
IC1	CA4011 or MC14011
IC2	CA4049 or MC14049
IC3	78L08 8-V regulator
Q1,2,3,4,6,9,10, 11,12,13	BS170, VN2222 or VN10 small power switch FET (pinout varies—see text)
Q5	MPSA13 Darlington pair
Q7,8	2N3904, 2N2222, etc BJT
Q14	IRF841, IRF820, IRF830, etc N-channel power FET
T Attenuator	
R1	33-Ω, 2-W carbon resistor (two parallel 68-Ω, 1-W carbon resistors)
R2	22-Ω, 1-W carbon resistor
R3	33-Ω, 1/4 or 1/2-W carbon resistor
p Attenuator:	*add jumper in place of R3*
R1	120-Ω, 1-W carbon resistor
RP2	75-Ω, 2 W carbon resistor (two parallel 150-Ω, 1-W carbon resistors)
RP3	120-Ω, 1/4 or 1/2-W carbon resistor
R4,R6	68-Ω, 1/4-W carbon resistor
R5	500-Ω small trimpot
R7,R8	10-Ω, 1/4-W resistor
R9	300-Ω, 1/4-W resistor
R10	5.6-kΩ, 1/4-W resistor
R11,14	10-kΩ, 1/4-W resistor
R12	2.7-kΩ, 1/4-W resistor
R13	560-Ω, 1/4-W resistor
R15	360-Ω, 1/4-W resistor
R16	430-Ω, 1/4-W resistor
R17	47-kΩ, 1/4-W resistor
R18	8.2-kΩ, 1/4-W resistor
R19	4.7-kΩ, 1/4-W resistor
R20	100-kΩ, 1/4-W resistor
R21,22,26	6.8-kΩ, 1/4-W resistor
R23,27	33-kΩ, 1/4-W resistor
R24	with LED: 1-kΩ, 1/4-W resistor; without LED: 0-Ω jumper (wire)
R25	with LED: 680-Ω, 1/4-W, resistor; without LED: 10-kΩ, 1/4-W resistor
RFC1,2	1-μH RF choke, molded
RLY1	DPDT relay, Radio Shack #275-249
Enclosure	Bud CU-124 or Hammond 1590B die-cast box

The 1 dB Quest Revisited

By John Swiniarski, K1OR and Bruce Wood, N2LIV

(From *Proceedings Of 1996 Microwave Update*)

Introduction

In 1992 Zack Lau, W1VT (ex-KH6CP), published an article[1] describing the design and construction of a 10.368 GHz HEMT low noise amplifier. Zack's work with the NEC 32684 lead to a very reproducible LNA that paved the way for sub 1 dB noise figure (NF) preamps used in many amateurs' 10 GHz stations.

Today, due to the proliferation of 12 GHz DBS systems, the quest for 1 dB NF has become commonplace. However, with the advances in devices offering lower and lower NF, many devices have become obsolete and discontinued such as the NEC 32684. Fortunately, with manufacturers striving for high volume manufacturing capability, this has led to less expensive devices.

We have investigated the use of some of the newer devices available by several manufacturers. Our primary goal was to determine the suitability of these devices for use with Zack's NE326 design. Secondly we hoped that some of these newer devices might offer improved performance over the NE326. Our methodology was to model the performance of the devices using Zack's original NE326 circuit. If the modeling indicated reasonable performance, preamps would be built on Zack's circuit board using identical components and construction techniques. These preamps were measured for NF and gain at the 22nd Eastern VHF/UHF Conference in Vernon, Connecticut. All devices tested seemed to perform well with no tuning or optimization. We anticipate attempting to tune these preamps in the near future.

Presented here is the outcome of the modeling study, some comments on construction and the measured results of the preamps. As can be seen, the quest for 1 dB can easily be achieved.

History

After many of the amateurs in the Northeast had received NEC 32684 HEMT devices as "door prizes" at local VHF conferences, considerable interest was raised in reproducing Zack's design. With the help of Down East Microwave, a quantity of printed circuit boards was made available to members of the North East Weak Signal (NEWS)

Group. Twenty-five or so preamps were constructed with the NE32684. It was then learned that NEC was replacing the '326 with the NE32584 that was shown to have a slightly lower NF. About the same time, a number of Fujitsu FHX 05 devices had become available. The FHX 05 is utilized in the Qualcomm 12 GHz receive LNAs and while it did not have as low a noise figure specification, its low cost made it an attractive candidate for study. In addition, an Avantek ATF 36077 was also made available for evaluation.

Modeling

EESOF Touchstone Software was utilized in the modeling study. Zack graciously provided a copy of his original circuit file. This eliminated the need to "reverse engineer" his circuit, an exercise that could lead to errors from making assumptions about the microstrip circuit elements. Having Zack's original model made the comparisons much more meaningful. A plot of the NF, Gain (S21) Input match (S11) and Output match (S22) of Zack's model is shown below:

The computed performance at 10.368 GHz is indicated by the markers on the chart. Using the same Touchstone circuit file, the S parameter and noise parameter file for the NE326 was replaced by the files for the NE325, FHX05 and the ATF360. Each new LNA was then "tuned" on the computer. This was done by changing the size and position of the open circuit stubs on the microstrip lines; much like we do with pieces of copper foil and a soldering iron on real hard-

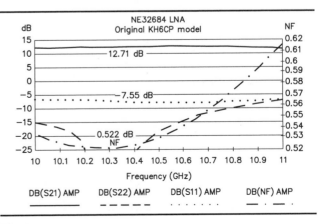

Fig 1

[1] Notes appear at the end of this section.

ware. The plots for the untuned condition and the corresponding tuned condition follows:

For visualization purposes, CAD drawings were made of the optimized LNAs. This might act as an aid when attempting optimization on the actual preamps. It can be seen that the changes are very subtle.

Construction

Samples of each of the devices were then built with identical connectors, capacitors, housings and bias circuits in an attempt to minimize variations. Assembly is fairly straight-forward, at least as far as microwave construction is concerned. Housings were fabricated using 0.500" by 0.025" hobby brass strips as described by Rus Healy, NJ2L.[2] This makes for a very compact device as the four hole SMA connectors used are also 0.500" high. The connectors used here were modified M/A-COM 2052-1215-00. This connector features a 0.085" diameter dielectric. The dielectric

length was shortened to match the 0.025" brass wall thickness. This leaves a 0.025" center contact which interfaces nicely with the microstrip PCB.

Several authors have described various methods for achieving source grounding for HEMT devices. These methods include bending the device leads through the board (difficult with most new devices as they are designed for surface mounting and therefore have short leads) and using rivets. Here, we simply made two parallel cuts about 0.050" long on either side of the device and used a "u" shaped piece of ribbon. The ribbon is pushed up through one slot and down the other then soldered to the ground plane. This results in a low inductance ground pad in which to solder the source leads.

Another unique feature is the miniature active bias board mounted below the RF printed circuit board. This board uses surface mount technology (SMT) devices to realize a circuit that can fit in the small area available. It is held in place by the copper straps that also provide grounding for

LNA Results

Computer Simulation

Device	Measured Result	Untuned Model	Tuned Mode
NE326#1	0.95 dB NF @ 11.35 dB gain	N/A	0.52 dB NF @ 12.71 dB gain
NE326#2	0.87 dB NF @ 11.39 dB gain	N/A	0.52 dB NF @ 12.71 dB gain
NE325#1	0.92 dB NF @ 13.45 dB gain	0.54 dB NF @ 13.28 dB gain	0.49 dB NF @ 15.02 dB gain
NE325#2	0.85 dB NF @ 13.20 dB gain	0.54 dB NF @ 13.28 dB gain	0.49 dB NF @ 15.02 dB gain.
ATF36077	0.95 dB NF @ 11.21 dB gain	0.55 dB NF @ 13.38 dB gain	0.52 dB NF @ 14.24 dB gain
FHX05	DOA	0.78 dB NF @ 10.89 dB gain	0.78 dB NF @ 12.98 dB gain

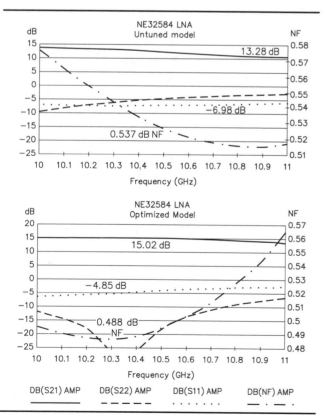

Fig 2

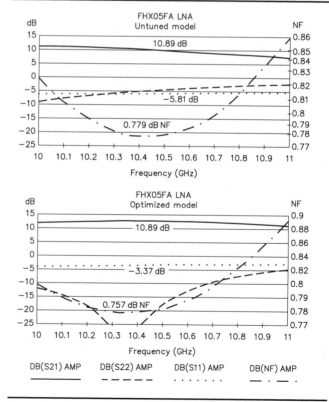

Fig 3

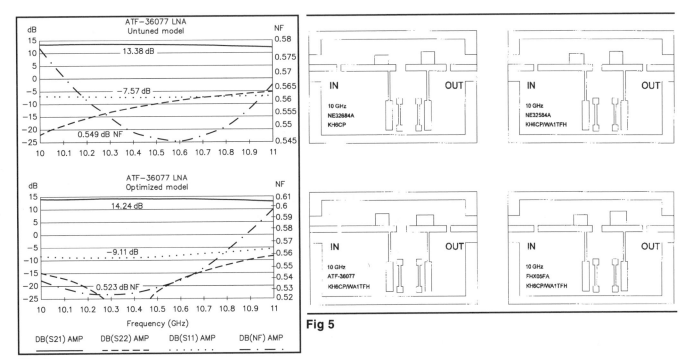

Fig 4

Fig 5

the board. The active bias circuit is similar to one described by Al Ward[3] and others. The –5 VDC supply is achieved using an ICL7660 voltage inverter chip. A spreadsheet was created to calculate the resistor values necessary for various bias conditions.

Results

The preamps were completed just in time for the 22nd Eastern VHF/UHF Conference in Vernon, Connecticut. Actually, several needed final wiring to the bias circuit, which was accomplished in the hotel room! The results obtained from the noise figure measuring session along with the computer simulation results are tabulated on the previous page. Unfortunately the Fujitsu FHX05 was DOA.

Conclusion

As the results here indicate, building a sub 1 dB NF preamp is readily achievable with several of today's hot devices. Each device tested resulted in less than 1 dB NF without optimization. The message appears to be "Use what you got." While the computer model seems to indicate much better NF performance should be possible, we are approaching the limits of what can be realized with readily available materials and construction techniques. We estimate at least 0.25 dB of loss exists ahead of the device. These are due to connector interfaces, component parasitics and dielectric losses. While it may be possible to tune these preamps to lower noise figures than measured, many builders of the original NE326 have been unsuccessful. The improvements indicated by the computer modeling of tuned vs. untuned performance hardly seem worth the trouble. Zack seems to have the magic, however. His original NE326 LNA always

10 Ghz LNA Assemblies

Fig 6

out performed all the others. His latest design based on the new NEC 329 device measured 0.70 dB NF at the conference. Perhaps his next quest should be for a 0.5 dB NF LNA for 10 GHz!

References:

[1]Lau, Z., KH6CP, "The Quest for 1 dB on 10 GHz," *QEX*, December 1992, pp. 16-19
[2]Healy, R., NJ2L, "Building Enclosures for Microwave Circuits," *QEX*, June 1994, pp. 15-17
[3]Ward, A. J., WB5LUA, "Simple Low-Noise Microwave Preamplifiers," *QST*, May 1989, pp. 31-36

Noise Measurement and Generation

Quality weak-signal reception requires a low-noise system. Here's how to calculate and measure the noise performance of your system

By Paul Wade, N1BWT

(From *QEX*, November 1996)

As anyone who has listened to a receiver suspects, everything in the universe generates noise. In communications, the goal is to maximize the desired signal in relation to the undesired noise we hear. To accomplish this goal, it would be helpful to understand where noise originates, how much our own receiver adds to the noise we hear, and how to minimize it.

It's difficult to improve something unless we can measure it. Measurement of noise in receivers does not seem to be clearly understood by many amateurs, so I will attempt to explain the concepts and clarify the techniques, and to describe the standard "measure of merit" for receiver noise performance: *noise figure*. Most important, I will describe how to build your own noise generator for noise-figure measurements.

A number of equations are included, but only a few are needed to perform noise-figure measurements. The rest are included as an aid to understanding, with, I hope, enough explanatory text for everyone.

Noise

The most pervasive source of noise is thermal noise, which arises from the motion of thermally agitated free electrons in a conductor. Since everything in the universe is at some temperature above absolute zero, every conductor must generate noise.

Every resistor (and all conductors have resistance) generates an RMS noise voltage:

$$e = \sqrt{4kTRB} \qquad \text{Eq 1}$$

where R is the resistance, T is the absolute temperature in kelvins (K), B is the bandwidth in hertz, and k is Boltzmann's constant, 1.38×10^{-23} joules/K.

Converting to power, $P = e^2/R$, and adjusting for the Gaussian distribution of noise voltage, the noise power generated by the resistor, in watts, is:

$$P_n = kTB \qquad \text{Eq 2}$$

which is independent of the resistance. Thus, all resistors at the same temperature generate the same noise power. The noise is white noise, meaning that the power density does not vary with frequency, but always has a power density of kT watts/Hz. The noise power is directly proportional to absolute temperature T, since k is a constant. At the nominal ambient temperature of 290 K, we can calculate this power; converted to dBm, we get the familiar -174 dBm/Hz. Just multiply by the bandwidth in hertz to get the available noise power at ambient temperature. The choice of 290 K for ambient might seem a bit cool, since the equivalent $17°$ C or $62°$ F would be a rather cool room temperature, but 290 K makes all the calculations come out to even numbers.

The *instantaneous* noise voltage has a Gaussian distribution around the RMS value. The Gaussian distribution has no limit on the peak amplitude, so at any instant the noise voltage may have any value from $-\infty$ to $+\infty$. For design purposes, we can use a value that will not be exceeded more than 0.01% of the time. This voltage is four times the RMS value, or 12 dB higher, so our system must be able to handle peak powers 12 dB higher than the average noise power if we are to measure noise without errors.[1]

Signal-to-Noise Ratio

Now that we know the noise power in a given bandwidth, we can easily calculate how much signal is required to achieve a desired signal-to-noise ratio (S/N). For SSB, perhaps 10 dB of S/N is required for good communication; the ambient thermal noise in a 2.5-kHz bandwidth is -140 dBm, calculated as follows:

$$P_n = kTB = 1.38 \times 10^{-23} \times 290 \times 2500 = 1.0 \times 10^{-17} \text{ W}$$

$$P_{dBm} = 10\log(P_n \times 1000) = -140 \text{ dBm}$$

(The factor of 1000 converts watts to milliwatts.) The signal power must be 10 dB greater than the noise power, so a minimum signal level of -130 dBm is required for a 10 dB S/N. This represents the noise and signal power levels at the antenna. We are then faced with the task of amplifying the signal without degrading the signal-to-noise ratio.

[1] Notes appear at the end of this section.

Noise Temperature

Any amplifier will add additional noise. The input noise N_i per unit bandwidth, kT_g, is amplified by gain G to produce an output noise of kT_gG. The additional noise added by the amplifier, kT_n is added to the input noise to produce a total noise output power N_o:

$$N_o = kT_gG + kT_n \qquad \text{Eq 3}$$

To simplify future calculations, we pretend that the amplifier is noise-free but has an additional noise-generating resistor of temperature T_e at the input, so that all sources of noise are inputs to the amplifier. Then the output noise is:

$$N_o = kG\left(T_g + T_e\right) \qquad \text{Eq 4}$$

and T_e is the *noise temperature* of the excess noise contributed by the amplifier. The noise added by an amplifier is then kGT_e, which is the fictitious noise generator at the input amplified by the amplifier gain.

Cascaded Amplifiers

If several amplifiers are cascaded, the output noise N_o of each becomes the input noise T_g to the next amplifier. We can create a large equation for the total. After removing the original input noise term, we are left with the added noise:

$$N_{added} = \left(kT_{e1}G_1G_2...G_N\right) + \left(kT_{e2}G_2...G_N\right) + ... + \left(kT_{eN}G_N\right) \qquad \text{Eq 5}$$

Substituting in the total gain $G_T = \left(G_1G_2...G_N\right)$ results in the total excess noise:

$$T_e = T_{e1} + T_{e2}/G_1 + T_{e3}/G_1G_2 + ... + T_{eN}/G_1G_2...G_{N-1} \qquad \text{Eq 6}$$

with the noise of each succeeding stage reduced by the gain of all preceding stages.

Clearly, if the gain of the first stage, G_1, is large, then the noise contributions of the succeeding stages are not significant. This is why we concentrate our efforts on improving the first amplifier or preamplifier.

Noise Figure

The noise figure (NF) of an amplifier is the logarithm of the ratio (so we can express it in dB) of the total noise output of an amplifier with an input T_g of 290 K to the noise output of an equivalent noise-free amplifier. A more useful definition is to calculate it from the excess temperature T_e:

$$NF = 10\log\left(1 + \frac{T_e}{T_0}\right) dB \text{ at } T_0 = 290 \text{ K} \qquad \text{Eq 7}$$

If the NF is known, T_e may be calculated after converting the NF to a ratio, F:

$$T_e = (F-1)T_0 = \left[10^{(NF/10)} - 1\right]T_0 \qquad \text{Eq 8}$$

Typically, T_e is specified for very low-noise amplifiers, where the NF would be a fraction of a dB, and NF is used when it seems a more manageable number than a T_e of thousands of kelvins.

Losses

We know that any loss or attenuation in a system reduces

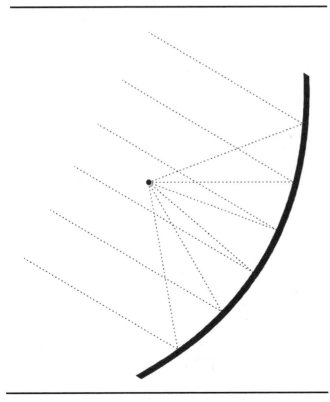

Fig 1—A parabolic dish aimed at a satellite.

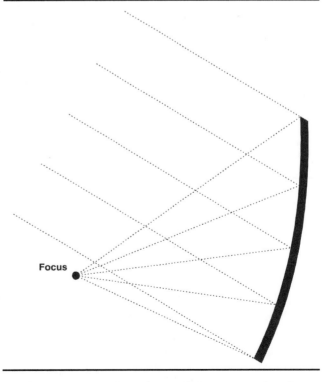

Fig 2—An offset-fed parabolic dish antenna aimed at a satellite.

the signal level. If attenuation also reduced the noise level, we could suppress thermal noise by adding attenuation.

Intuitively, this can't be true. The reason is that the attenuator—or any lossy element—has a physical temperature, T_x, that contributes noise to the system while the input noise is being attenuated. The output noise after a loss L (expressed as a ratio, not in dB) is:

$$T_{g'} = \frac{T_g}{L} + \left(\frac{L-1}{L}\right) T_x \qquad \text{Eq 9}$$

If the source temperature T_g is higher than the attenuator temperature T_x, the noise contribution is the familiar result found by simply adding the loss in dB to the NF. However, for low source temperatures the degradation can be much more dramatic. If we do a calculation for the effect of 1 dB of loss (L = 1.26) on a T_g of 25 K:

$$T_{g'} = \frac{25}{1.26} + \left(\frac{0.26}{1.26}\right) \times 290 = 80 \text{ K}$$

The resultant $T_{g'}$ is 80 K, *a 5 dB increase in noise power* (or a 5-dB degradation of signal-to-noise ratio). Since noise power = kT and k is a constant, the increase is the ratio of the two temperatures 80/25, or in dB, 10 log (80/25)=5 dB.

Antenna Temperature

How can we have a source temperature much lower than ambient? If an antenna, assumed to be lossless, is receiving signals from space, rather than the warm earth, then the background noise is much lower. The background temperature of the universe has been measured as about 3.2 K. An empirical number for a 10-GHz antenna pointing into clear sky is about 6 K, since we must always look through the attenuation and temperature of the atmosphere.[2] The figure will vary with frequency, but a good EME antenna might have a T_g of around 20 K at UHF and higher frequencies.

A couple of examples of actual antennas might bring all of this together.[3]

1. A 30-inch conventional dish at 10 GHz, with a measured gain of 36.4 dBi and efficiency of 64%. The estimated spillover efficiency is 87% for a 10-dB illumination taper. With the dish pointing at a high elevation, as shown in Fig 1, perhaps half of the spillover is illuminating Earth at 290 K, which adds an estimated 19 K to the 6 K of sky noise, for a total of 25 K. In a 500-Hz bandwidth, the noise output is –157.6 dBm.

2. An 18-inch DSS offset-fed dish at 10 GHz, with measured gain of 32.0 dB and efficiency of 63%. The spillover efficiency should be comparable, but with the offset dish pointing at a high elevation, as shown in Fig 2, far less of the spillover is illuminating warm Earth. If we estimate 20%, then 8 K is added to the 6 K of sky noise, for a total of 14 K. In a 500-Hz bandwidth, the noise output is –160 dBm.

The larger conventional dish has 2.4 dB higher noise output but 4.4 dB higher gain, so it should have 2.0 dB better signal-to-noise ratio than the smaller offset dish when both are pointing at high elevations.

However, while the offset dish is easy to feed with low loss, it is more convenient to feed the conventional dish through a cable with 1 dB of loss. Referring back to our loss example above, the noise temperature after this cable loss is 80 K. In a 500-Hz bandwidth, the noise output is now –152.6 dBm, 7.4 dB worse than the offset dish. The convenience of the cable reduces the signal-to-noise ratio by 5 dB, making the larger conventional dish 3 dB worse than the smaller offset dish. Is it any wonder that the DSS dishes sprouting on rooftops everywhere are offset-fed?

If the dishes are pointed at the horizon for terrestrial operation, the situation is much different. At least half of each antenna pattern is illuminating warm Earth, so we should expect the noise temperature to be at least half of 290 K, or about 150 K. Adding 1 dB of loss increases the noise temperature to 179 K, a 1 dB increase. At higher noise temperatures, losses do not have a dramatic effect on signal-to-noise ratio. In practice, the antenna temperature on the horizon may be even higher since the upper half of the pattern must take a much longer path through the warm atmosphere, which adds noise just like any other loss.

Image Response

Most receiving systems use at least one frequency-converting mixer that has two responses, the desired frequency and an image frequency on the other side of the local oscillator. If the image response is not filtered out, it will add additional noise to the mixer output. Since most preamps are broadband enough to have significant gain (and thus, noise output) at the image frequency, the filter must be placed between the preamp and the mixer. The total NF including image response is calculated:

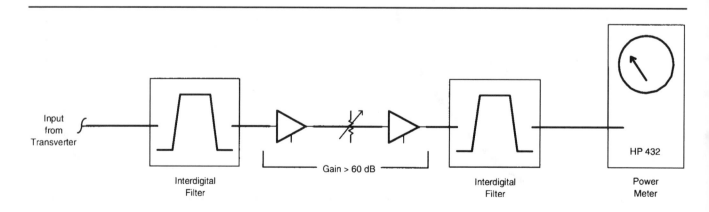

Fig 3—System for measuring sun noise.

$$NF = 10\log\left[\left(\frac{1+T_e}{T_0}\right)\left(1+\frac{G_{image}}{G_{desired}}\right)\right] \qquad \text{Eq 10}$$

assuming equal noise bandwidth for desired and image responses. Without any filtering, $G_{image} = G_{desired}$ so $1+\left(G_{image}/G_{desired}\right) = 2$, doubling the noise figure, which is the same as adding 3 dB. Thus, without any image rejection, the overall noise figure is at least 3 dB *regardless of the NF of the preamp*. For the image to add less than 0.1 dB to the overall NF, a quick calculation shows that the gain at the image frequency must be at least 16 dB lower than at the operating frequency.

Noise Figure Measurement

So far we have discussed the sources of noise and a figure of merit for evaluating the receiving system's response to noise. How can we measure an actual receiver?

The noise figure of a receiver is determined by measuring its output with two different noise levels, T_{hot} and T_{cold}, applied to its input. The ratio of the two output levels is referred to as the *Y-factor*. Usually, the ratio is determined from the difference in dB between the two output levels, Y_{dB}:

$$Y_{(ratio)} = 10^{\left(Y_{dB}/10\right)} \qquad \text{Eq 11}$$

Then the receiver T_e may be calculated using $Y_{(ratio)}$:

$$T_e = \frac{T_{hot} - Y \times T_{cold}}{Y-1} \qquad \text{Eq 12}$$

and converted to noise figure:

$$NF = 10\log\left(1+\frac{T_e}{T_0}\right)dB \qquad \text{Eq 13}$$

where $T_0 = 290$ K

The two different noise levels may be generated separately, for instance by connecting resistors at two different temperatures. Alternatively, we could use a device that can generate a calibrated amount of noise when it is turned on. When such

a device is turned off, it still generates noise from its internal resistance at T_{cold}, the ambient temperature (290 K); usually this resistance is 50 Ω, to properly terminate the transmission line that connects it to the receiver. When the noise generator is turned on, it produces excess noise equivalent to a resistor at some higher temperature at T_{hot}. The noise produced by a noise source may be specified as the Excess Noise Ratio (ENR_{dB}), the dB difference between the cold and the equivalent hot temperature, or as the equivalent temperature of the excess noise, T_{ex}, which is used in place of T_{hot} in Eq 12. If the ENR is specified, then the calculation is:

$$NF_{dB} = ENR_{dB} - 10\log\left(Y_{(ratio)}-1\right) \qquad \text{Eq 14}$$

The terms T_{ex} and ENR are used rather loosely; assume that a noise source specified in dB refers to ENR_{dB}, while a specification in "degrees" or kelvins refers to T_{ex}.

An automatic noise-figure meter, sometimes called a PANFI (*precision automatic noise-figure meter*), turns the noise source on and off at a rate of about 400 Hz and performs the above calculation electronically.[4] A wide bandwidth is required to detect enough noise to operate at this rate; a manual measurement using a narrow-band communications receiver would require the switching rate to be less than 1 Hz, with some kind of electronic integration to properly average the Gaussian noise.

Noise-figure meters seem to be fairly common surplus items. The only one in current production, the HP 8970, measures both noise figure and gain but commands a stiff price.

AIL (later AILTECH or Eaton) made several models; the model 2075 measures both NF and gain, while other models are NF only. The model 75 (a whole series whose model numbers start with 75) shows up frequently for anywhere from $7 to $400, typically $25 to $50, and performs well. Every VHFer I know has one, with most of them waiting for a noise source to be usable. Earlier tube models, like the AIL 74 and the HP 340 and 342, have problems with drift and heat, but they can also do the job.

Another alternative is to build a noise-figure meter.[5]

Fig 4—A noise source built on an SMA connector.

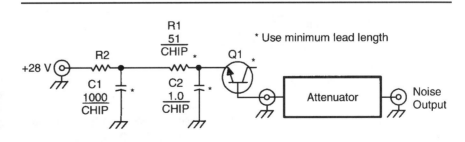

Fig 5—Schematic diagram of a noise source built on an SMA connector.

Q1—Tiny silicon NPN RF transistor such as NEC 68119.

R2—Select to set current (see text). 1 kΩ minimum, ¼ W.

Using the Noise-Figure Meter

I'll describe the basic procedure using the model 75; others are similar, but the more complex instruments will require studying the instruction manual.

Input to almost all noise-figure meters is at 30 MHz, so a frequency converter is required (some instruments have internal frequency converters; except for the HP 8970, I'd avoid using this feature). Most ham converters with a 28-MHz IF work fine, unless the preamp being measured is so narrowband that a megahertz or two changes the NF. The input is fairly broadband, so LO leakage or any other stray signals can upset the measurement—this has been a source of frustration for many users. There are two solutions: a filter (30-MHz low-pass TVI filters are often sufficient) or a tuned amplifier at 30 MHz. Since a fair amount of gain is required in front of the noise figure meter, an amplifier is usually required anyway.

A noise source (which we will discuss in detail later) is connected to the rear of the instrument: a BNC connector marked DIODE GATE provides +28 V for a solid-state noise source, and high-voltage leads for a gas-tube noise source are also available on many versions. The noise-figure meter switches the noise source on and off. The noise output coax connector of the noise source is connected to the receiver input.

The model 75 has four function pushbuttons: OFF, ON, AUTO, and CAL. The OFF and ON positions are for manual measurements: OFF displays the detector output with the noise source turned off, and ON displays the detector output with the noise source turned on. If all is working, there should be more output in the ON position, and a step attenuator in the IF line may be used to determine the change in output, or Y-factor, to sanity-check our results. The knob marked GAIN is used to get the meter reading to a desirable part of the scale in the OFF and ON positions only; it has no effect on automatic measurements.

The AUTO position causes the instrument to turn the noise source on and off at about a 400-Hz rate and to calculate the NF from the detected change in noise. The model 75 has a large green light near the meter which indicates that the input level is high enough for proper operation—add gain until the light comes on. Then the meter should indicate a noise figure, but not a meaningful one, since we must first set the ENR$_{dB}$ using the CAL position. The lower scale on the meter is marked for from 14.5 to 16.5 dB of ENR; adjust the CAL ADJ knob until the reading in the CAL position matches the ENR of the noise source.

If the ENR of your noise source is outside the marked range, read the section below on homebrew noise sources.

Now that we have calibrated the meter for the ENR of the noise source, we may read the noise figure directly in the AUTO position. Before we believe it, a few sanity checks are in order:

• Manually measure and calculate the Y-factor.

• Insert a known attenuator between the noise source and preamp—the NF should increase by exactly the attenuation added.

• Measure something with a known noise figure (known means measured elsewhere; a manufacturer's claim is not necessarily enough).

Finally, too much gain in the system may also cause trouble if the total noise power exceeds the level that an amplifier stage can handle without gain compression. Gain compression will be greater in the on state, so the detected Y-factor will be reduced, resulting in erroneously high indicated NF. The Gaussian distribution of the noise means that an amplifier must be able to handle 12 dB more than the average noise level without compression. One case where this is a problem is with a microwave transverter to a VHF or UHF IF followed by another converter to the 30-MHz noise-figure meter, for too much total gain. I always place a step attenuator between the transverter and the converter,

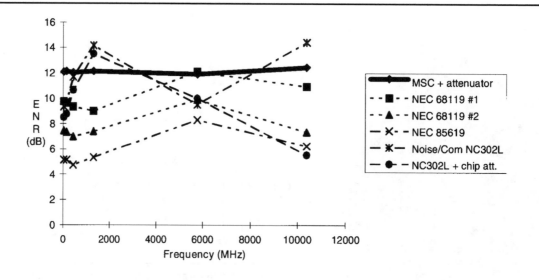

Fig 6—ENR of several versions of the noise source of Fig 5.

which I adjust until I can both add and subtract attenuation without changing the indicated noise figure.

One final precaution: noise-figure meters have a very slow time constant, as long as 10 seconds for some of the older models, to smooth out the random nature of noise. If you are using the noise-figure meter to "tweak" a receiver, *tune very slowly!*

Sky-Noise Measurement

Another way to measure noise figure at microwave frequencies is by measurement of sky noise and ground noise.[3,6] Sky noise is very low, around 6 K at 10 GHz, for instance, and ground noise is due to the ground temperature, around 290 K, so the difference is nearly 290 K. At microwave frequencies we can use a manageable antenna that is sharp enough that almost no ground noise is received, even in sidelobes, when the antenna is pointed at a high elevation. A long horn would be a good antenna choice.

The antenna is pointed alternately at clear sky overhead, away from the sun or any obstruction, and at the ground. The difference in noise output is the Y-factor; since we know both noise temperatures, the receiver noise temperature is calculated using the $Y_{(ratio)}$ and Eq 12.

The latest version of my microwave antenna program, *HDLANT21*, will make this calculation.[3] Since the measured Y-factor will be relatively small, this measurement will only be accurate for relatively low noise figures. On the other hand, they are the most difficult to measure accurately using other techniques.

A system for measuring sun noise was described by Charlie, G3WDG, that also works well for measuring noise figure from sky noise.[7] He built a 144-MHz amplifier with moderate bandwidth using MMICs and helical filters that amplifies the transverter output to drive a surplus RF power meter. The newer solid-state power meters are stable enough to detect and display small changes in noise level,

and the response is slow enough to smooth out flicker. Since my 10-GHz system has an IF output at 432 MHz, duplicating Charlie's amplifier would not work. In the junk box I found some surplus broadband amplifiers and a couple of interdigital filters, which I combined to provide high gain with bandwidth of a few megahertz, arranged as shown in Fig 3. I found that roughly, 60 dB of gain after the transverter was required to get a reasonable level on the power meter, while the G3WDG system has somewhat narrower bandwidth so more gain is required.

Several precautions are necessary:

• Peak noise power must not exceed the level that any amplifier stage can handle without gain compression. Amplifiers with broadband noise output suffer gain compression at levels lower than found with signals, so be sure the amplifier compression point is at least 12 dB higher than the indicated average noise power.

• Make sure no stray signals appear within the filter passband.

• Foliage and other obstructions add thermal noise that obscures the cold sky reading.

• Low-noise amplifiers are typically very sensitive to input mismatch, so the antenna must present a low VSWR to the preamp.

A noise-figure meter could also be used as the indicator for the sky-noise measurement, but a calibrated attenuator would be needed to determine the Y-factor. Using different equipment gives us an independent check of noise figure so we may have more confidence in our measurements.

W2IMU suggested that the same technique could be used for a large dish at lower frequencies.[8] With the dish pointing at clear sky, the feed horn is pointing at the reflector, which shields it from the ground noise so it only sees the sky noise. If the feed horn is then removed and pointed at the ground, it will see the ground noise.

Noise-figure meters are convenient, but if you don't have one, the equipment for measuring sun and sky noise could also be used indoors with a noise source. The only complication is that the Y-factor could be much larger, pushing the limits of amplifier and power meter dynamic range.

Noise Sources

The simplest noise source is simply a heated resistor— if we know the temperature of the resistor, we can calculate exactly how much noise it is generating. If we then change

Fig 7—The homebrew noise source of Fig 8.

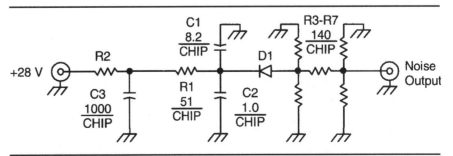

Fig 8—Schematic diagram of the homebrew noise source using a chip-resistor attenuator. C1 and C2 are microwave chip capacitors, ATC or equivalent.

the temperature, the noise output will change by a known amount. This would work if we could find a resistor with good RF properties whose value does not change with temperature, an unlikely combination. There are commercial units, called *hot-cold noise sources*, with two calibrated resistors at different temperatures and low VSWR. Typically, one resistor is cooled by liquid nitrogen to 77.3 K (the boiling point of nitrogen), while the other is heated by boiling water to 100° C, or 373.2 K. The preamp is connected to first one resistor, then the other; the difference in noise output is the Y-factor.

Since the boiling point of pure liquids is accurately known, this type of noise generator can provide very accurate measurements. However, they are inconvenient to use, since the receiver must be connected directly to alternate resistors (the loss in an RF switch would significantly reduce the noise output and accuracy). Also, few amateurs have a convenient source of liquid nitrogen.

Three types of noise sources are commonly available and convenient to use:

1. Temperature-limited vacuum tube diode. The noise output is controlled by the diode current but is only accurate up to around 300 MHz due to limitations of the vacuum tube. These units generate around 5 dB of excess noise.

2. Gas tube sources. The noise is generated by an ionized gas in the tube, similar to a fluorescent light—homebrew units have been built using small fluorescent tubes. The noise tubes use a pure gas, typically argon, to control the noise level. These units typically generate about 15 dB of excess noise.

Coaxial gas tube sources work up to around 2.5 GHz, and waveguide units to much higher frequencies. One problem using these is that a high voltage pulse is used to start the ionization (like the starter in a fluorescent light) which is coupled to the output in the coaxial units and is large enough to damage low-noise transistors. Since a noise-figure meter turns the noise source on and off continuously, pulses are generated at the same rate.

Since waveguide acts as a high-pass filter, the starting pulses are not propagated to the output, so wave-guide gas-tube noise sources are safe to use, though bulky and inconvenient. However, they could be used to calibrate a solid-state noise source.

Another problem with all gas tubes is that the VSWR

of the noise source changes between the on and off states. If the source VSWR changes the noise figure of an amplifier, as is almost always the case, the accuracy of the measurement is reduced.

3. Solid-state noise sources. Reverse breakdown of a silicon diode PN junction causes an avalanche of current in the junction that would rise to destructively high levels if not limited by an external resistance. Since current is "electrons in motion," a large amount of noise is generated. If the current density of the diode is constant, the average noise output should also be constant; the instantaneous current is still random with a Gaussian distribution, so the generated noise is identical to thermal noise at a high temperature. Commercial units use special diodes designed for avalanche operation with very small capacitance for high-frequency operation, but it is possible to make a very good noise source using the emitter-base junction of a small microwave transistor.

Typical noise output from an avalanche noise diode is 25 dB or more, so the output must be reduced to a usable level, frequently 15 dB of excess noise to be compatible with gas tubes or 5 dB of excess noise for more modern equipment. If the noise level is reduced by a good RF attenuator of 10 dB or more, the source VSWR (seen by the receiver) is dominated by the attenuator, since the minimum return loss is twice the attenuation. Thus, the change in VSWR as the noise diode is turned on and off is minuscule. Commercial noise sources consist of a noise diode assembly and a selected coaxial attenuator permanently joined in a metal housing, calibrated as a single unit.

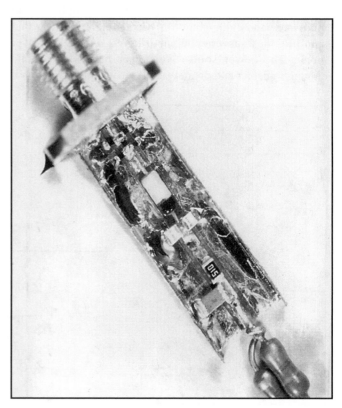

Fig 10—The noise source of Fig 8, constructed on a photographically printed circuit board.

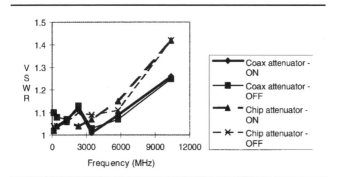

Fig 9—Measured VSWR of homebrew noise sources of Figs 4 and 7.

Homebrew Noise Sources

There are three components of a noise source: a noise generator, an attenuator and the calibration data—the ENR at each frequency. The most critical component is the attenuator; it is very important that the noise source present a very low VSWR to the preamp or whatever is being measured since low-noise amplifiers are sensitive to input impedance, and even more important that the VSWR does not change significantly when the noise source is turned on and off since a change causes error in the measurement. Because an attenuator provides twice as many dB of isolation as loss (reflections pass through a second time), 10 dB or more of attenuation will reduce any change in VSWR to a very small value.

Commercial solid-state noise sources occasionally appear in surplus sources, usually at high prices but occasionally very cheap if no one knows what it is. I have found two of the latter, and one of them works! It produces about 25 dB of excess noise, which is too much to be usable. I went through my box of hamfest attenuators and found one that has excellent VSWR up to 10 GHz and 13 dB of attenuation. Mated with the noise source, the combination produces about 12 dB of excess noise—a very usable amount. Finally, I calibrated it against a calibrated noise source for all ham bands between 50 MHz and 10 GHz; not exactly NTIS traceable, but pretty good for amateur work.

While noise sources are hard to locate, noise-figure meters are frequent finds. If we could come up with some noise sources, all the VHFers who have one gathering dust could be measuring and optimizing their noise figure.

Several articles have described construction of homebrew noise sources that work well at VHF and UHF but not as well at 10 GHz.[9,10,11] All of them have the diode in a shunt configuration, with one end of the diode grounded. When I disassembled my defective commercial noise source (even the attenuator was bad), I found a bare chip diode in a series configuration—diode current flows into the output attenuator. Obviously I could not repair a chip diode, but I could try the series diode configuration. I found the smallest packaged microwave transistor available, some small chip resistors and capacitors, and soldered them directly on the gold-plated flange of an SMA connector with zero lead length, as shown in the photograph, Fig 4. We've all soldered components directly together in "dead-bug" construc-

tion; this is more like "fly-speck" construction. The schematic is shown in Fig 5, and it works at 10 GHz! I built several versions to evaluate reproducibility and measured them at several ham bands from 30 MHz to 10 GHz, with the results shown in Fig 6. All units were measured with the same 14-dB attenuator, so the diode noise generator output is 14 dB higher.

(Later I found that the MIT Radiation Laboratory had described a noise source with a series diode 50 years ago, so we aren't giving away anyone's trade secrets.)[12]

I then remembered that I had a commercial noise diode, a Noise/Com NC302L, which was used in a noise source described in *QST*, with the diode in the shunt configuration.[11] The diode is rated as working to 3 GHz, so, in the amateur tradition, I wanted to see if I could push it higher, using the series configuration. Since I didn't expect to reach 10 GHz, I increased the value of the bypass capacitor, but otherwise, it looks like the units in Fig 5. When I measured this unit, it not only worked at 10 GHz, but had more excess noise output than at lower frequencies, probably due to an unexpected resonance. The performance is shown in Fig 6 along with the other units.

Also shown in Fig 6 is the output of my pseudo-commercial noise source; even with the external attenuator, the excess noise output is pretty flat with frequency. Commercial units are typically specified at ±0.5 dB flatness. In Fig 6, none of the homebrew ones are that flat, but there is no need for it; as long as we know the excess noise output for a particular ham band, it is perfectly usable for that band.

All the above noise sources rely on a coaxial microwave attenuator to control the VSWR of the noise source. Attenuators are fairly frequent hamfest finds, but ones that themselves have good VSWR to 10 GHz are less common, and it's hard to tell how good they are without test equipment. An alternative might be to build an attenuator from small chip resistors. I used my PAD.EXE program to review possible resistor values, and found that I could make a 15.3-dB π attenuator using only 140-Ω resistors if the shunt legs were formed by two resistors in parallel, a good idea to reduce stray inductance.[13] I ordered some 0402-size (truly tiny) chip resistors from Digi-Key and more NC302L diodes from Noise/Com, and built the noise source shown in Fig 7 on a bit of Teflon PC board, cutting out the 50-Ω transmission line with a hobby knife. The schematic of the

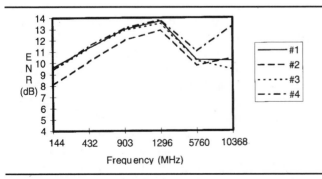

Fig 11—ENR of the PC-board noise sources (Fig 10).

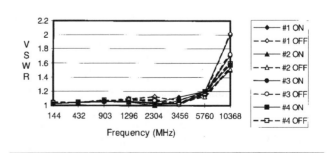

Fig 12—Measured VSWR of the PC-board noise sources (Fig 10).

Enhance Miniature Construction with Optical Feedback

Many microwave construction projects, and most modern equipment, use extremely small surface-mount components. Working with these parts requires steady hands and good vision. And as we get older, our vision usually deteriorates—I got my first bifocals last year.

I'm convinced that the key to working with tiny parts is to see them well. When I built the first noise source with the tiny 0402-size (1 × 0.5 mm) chip resistors, it was frustrating trying to get the resistors soldered where I wanted them. After that experience, I was on the lookout for a surplus stereo microscope and finally located one at a reasonable price. These microscopes are commonly used for microelectronics assembly work, providing moderate magnification at a long working distance.

After I set up the microscope on my workbench with adequate illumination, I was ready to build some more noise sources. Now the tiny chip resistors were clearly visible, and I was able to hold them in place with tweezers while soldering them exactly where I wanted them. On another project, I wanted a clearance hole in the ground plane around a hole drilled though a Teflon PC board. The hole is 0.025 inches in diameter, and I was able to cut an octagon around it with a hobby knife; the length of each side of the octagon is about the same as the hole diameter. Then I lifted the unwanted copper with the point of the knife. Magnification makes miniature work feel precise and easy instead of clumsy and frustrating.

What the microscope does is add gain to the feedback loop from the eyes to the hand. Our hands are never perfectly steady, but adding this feedback steadies them under the microscope, as the brain takes input from the eyes and automatically compensates (after a bit of practice).

A microscope is an elegant solution for very small parts, but any optical magnification helps. I have also used magnifiers, jewelers loupes and "drugstore" reading glasses. If the reading glasses are stronger than you need, they will provide additional magnification; just don't try walking around wearing them.

Other aids to miniature work are tweezers, fine-point soldering irons and lots of light. When an object is magnified, proportionally more light is required for the same apparent brightness. Tweezers help in holding small objects—I prefer the curved #7 style Swiss tweezers, of stainless steel so solder won't stick. Finally, a temperature-controlled soldering iron prevents overheating, which can destroy the solder pads on surface-mount components; 700° F tips are hot enough. All the tools I've mentioned came from hamfests, surplus places or flea markets, at reasonable prices.

So, even if you think that microwave project with tiny parts is beyond your capability, use a magnifier and give it a try. I'll bet you surprise yourself.—*N1BWT*

complete noise source is shown in Fig 8.

The chip-resistor attenuator works nearly as well as an expensive coaxial one. The measured VSWR of two noise sources, one with the chip attenuator and the other with a coaxial attenuator, is shown in Fig 9. Curves are shown in both the off and on states, showing how little the VSWR changes. The VSWR of the chip attenuator unit is 1.42 at 10 GHz, slightly over the 1.35 maximum specified for commercial noise sources, but still fine for amateur use.

Still, I wondered if I could do even better. The hand-cut board used is 0.031-inch-thick Teflon material, which is a bit thick at 10 GHz. I obtained some 0.015-inch-thick material and made a photo mask to print an accurate 50-Ω line. Then I carefully assembled the components under a microscope (see sidebar: Enhance Miniature Construction with Optical Feedback). Fig 10 shows the construction: the thin PC board is supported by a thin brass strip soldered along each side to create miniature I-beam, a much sturdier structure. The brass strips also connect the top and bottom ground areas of the board.

I built four units like the one shown in Fig 10, with encouraging results. The ENR of these units, shown in Fig 11, was higher at 10 GHz than the hand-cut one and reasonably flat with frequency—and consistent from unit to unit. The VSWR, however, was still high at 10 GHz, as shown in Fig 12. It is difficult to make a really good coax-to-microstrip transition at 10 GHz! Ordinarily, in an amplifier, we simply tune out the slight mismatch as part of the tuning procedure, but broadband tuning is much more difficult. As a final improvement, I dug up some 5-dB SMA attenuators from the swap session at Microwave Update last year. Adding one of these to the worst unit in Fig 12 reduced the VSWR to 1.18 at 10 GHz (below 1.10 at lower frequencies) and the ENR to 5.0 dB. This performance is every bit as good as a very expensive commercial noise source, lacking only NTIS-traceable calibration.

Noise-Source Alignment

The only alignment requirement for a solid-state noise source is to set the diode current; the current is always set at the highest frequency of interest. A noise figure meter must be set up with converters, etc, for the highest frequency at which the noise source might be used and set to display the detector output (OFF position on a model 75). Then voltage from a variable dc power supply is applied to the noise diode through the 1-kΩ current-limiting resistor. The detector output should increase as the voltage (diode current) increases, reach a peak, then decrease slightly. The optimum current is the one that produces peak output at the highest frequency (I set mine at 10 GHz). Then additional resistance must be added in series with the current-limiting resistor so that the peak output occurs with 28 volts applied, so that the noise source may be driven by the noise-figure meter. Once the proper resistor is determined and added, the dc end of the noise source is connected to the diode output of the noise-figure meter and the meter function is set to ON. This should produce the same detector output as the power supply.

Then the meter function is set to AUTO and the meter

should produce some noise-figure indication, but not yet a calibrated one. However, it is good enough to tune up preamps—a lower noise figure is always better, even if you don't know how low it is.

Noise-Source Calibration

Much of the high price of commercial noise sources pays for the NTIS-traceable calibration. Building a noise source only solves part of the problem—now we need to calibrate it.

The basic calibration technique is to measure something with a known noise figure using the new noise source, then calculate what ENR would produce the indicated noise figure.

Fortunately, the calculation is a simple one involving only addition and subtraction; no fancy computer program required. Simply subtract the indicated noise figure, $NF_{indicated}$, from the known noise figure, NF_{actual}, and add the difference to the ENR for which the meter was calibrated, ENR_{cal}:

$$ENR_{(noise\ source)} = ENR_{cal} + (NF_{actual} - NF_{indicated}) \qquad Eq\ 15$$

This procedure must be repeated at each frequency of interest; at least once for each ham band should be fine for amateur use.

The known noise figure is best found by making the measurement with a calibrated noise source, then substituting the new noise source so there is little opportunity for anything to change. Next best would be a sky noise measurement on a preamp. Least accurate would be to measure a preamp at a VHF conference or other remote location, then bring it home and measure it, hoping that nothing rattled loose on the way. If you can't borrow a calibrated noise source, it would be better to take your noise source elsewhere and calibrate it. Perhaps we could measure noise sources as well as preamps at some of these events.

Using the Noise Source

Now that the ENR of the noise source has been calibrated, the noise-figure meter calibration must be adjusted to match. However, the model 75 in the CAL position has only 2 dB of adjustment range marked on the meter scale. Older instruments have no adjustment at all. However, we can just turn around the equation we used to calculate the ENR and calculate the NF instead:

$$NF_{actual} = NF_{indicated} + \left(ENR_{(noise\ source)} - ENR_{cal} \right) \qquad Eq\ 16$$

There is a short cut. My noise source has an ENR around 12 dB, so I set the CAL ADJ in the CAL position as if the ENR were exactly 3 dB higher, then subtract 3 dB from the reading. Even easier, the meter has a +3 dB position on the ADD TO NOISE FIGURE switch. Using that position, I can read the meter starting at 0 dB. Any ENR difference from 15 dB that matches one of the meter scales would also

work—rather than an involved explanation, I'd urge you to do the noise figure calculations, then try the switch positions and see what works best for quick readout.

Reminder: Noise-figure meters have a very slow time constant, as long as 10 seconds for some of the older models, to smooth out the random nature of noise. *Tune slowly!*

Don't despair if the ENR of your noise source is much less than 15 dB. The optimum ENR is about 1.5 dB higher than the noise figure being measured.[1] The fact that today's solid-state noise sources have an ENR around 5 dB rather than the 15 dB of 20 years ago shows how much receivers have improved.

Conclusion

The value of noise-figure measurement capability is to help us all to "hear" better. A good noise source is an essential part of this capability. Accurate calibration is not necessary but helps us to know whether our receivers are as good as they could be.

Notes
[1]Pettai, R., *Noise in Receiving Systems*, Wiley, 1984.
[2]Graves, M. B., WRØI, "Computerized Radio Star Calibration Program," *Proceedings of the 27th Conference of the Central States VHF Society*, ARRL, 1993, pp 19-25.
[3]Wade, P., N1BWT, "More on Parabolic Dish Antennas," *QEX*, December 1995, pp 14-22. The *HDLANT21* program may be downloaded from **http://www.arrl.org/qexfiles**.
[4]Pastori, W. E., "Direct-Reading Measurement of Receiver-Noise Parameters," *Microwave Journal*, April 1973, pp 17-22.
[5]Bertelsmeier, R., DJ9BV, and Fischer, H., DF7VX, "Construction of a Precision Noise Figure Measuring System," *DUBUS Technik 3*, DUBUS, 1992, pp 106-144.
[6]Fasching, H., OE5JFL, "Noise Figure Measurement using Standard Antennas," *DUBUS Technik 4*, DUBUS, 1995, pp 23-25.
[7]Suckling, C., G3WDG, "144 MHz wideband noise amplifier," *DUBUS 2/1995*, DUBUS, 1995 pp 5-8.
[8]Turrin, R., W2IMU, "Method for Estimating Receiver Noise Temperature," Crawford Hill Technical Note #20, September 1986.
[9]Britain, K., WA5VJB, "10 GHz Noise Source," *Microwave Update '87*, ARRL, 1987, p 63.
[10]Wade, P., N1BWT, and Horsefield, S., NR1E, "Homebrew Solid State Noise Sources," *Proceedings of the 1992 (18th) Eastern VHF/UHF Conference*, ARRL, 1992.
[11]Sabin, W. E., WØIYH, "A Calibrated Noise Source for Amateur Radio," *QST*, May 1994, pp 37-40.
[12]Valley, G. E., Jr., and Wallman, H., *Vacuum Tube Amplifiers*, MIT Radiation Laboratory Series, McGraw-Hill, 1948.
[13]Wade, P., N1BWT, "Building VHF Power Attenuators," *QEX*, April 1994, pp. 28-29. The PAD.EXE program may be downloaded from **http://www.arrl.org/qexfiles**.

More Pipe-Cap Mania

By John Sortor, KB3XG

(From *QEX*, April 1996)

I needed a band-pass filter to reduce the noise bandwidth of my 3456-MHz receiver lineup. I tried the "beg, borrow or steal" method with no success. Bill, W3HQT, gave me a copy of a 5760-MHz transverter article which used pipe-cap filters on the mixers. I scaled the dimensions for a 3456 version and it just so happens that a 1¹/₂-inch pipe cap works fine.

I used a 2-inch square piece of 0.063-inch G-10 fiberglass board, two pieces of 0.085-inch semi-rigid coax with SMAs, a 1¹/₂-inch 10-32 brass screw and two brass nuts. The probe spacing equals ¹/₄ (0.854 inch at 3456 MHz) and the probe length equals ¹/₈ (0.427 inch at 3456 MHz).

The semi-rigid shield should protrude ¹/₁₆ inch into the cavity side. Solder the coax on both sides of the PC board for mechanical strength. Measure the length of the probes after completing the soldering process. The heat may cause the Teflon to walk out of the copper jacket. Drill a hole in the top of the pipe cap and solder a brass nut to the cap. Use the other nut as a jam nut to assure a good electrical contact to the pipe cap.

Secure the pipe cap to the PC board using two small C-clamps. Connect your Down East no-tune transverter or equivalent to the filter. Move the pipe cap around to minimize the filter loss. Simultaneously re-tweak the tuning screw to find the optimum spot for the shell and screw. Tighten the C-clamps and solder the pipe cap to the PC board.

I can't believe this thing works as well as it does. I got about 1 dB of loss at 3456 MHz with a 3-dB bandwidth of about 50 MHz. The filter dropped off rapidly below 3.4 GHz, but flat topped above 3.4 GHz. There was another peak at 7 GHz where the probes are ¹/₄ waves. This works fine for us since we are trying to reject our LO signal. WA3NUF says that shortening the probes will increase the loss but sharpen the filter skirts. Feel free to experiment at your own expense. The pipe caps are two bucks a piece, so you don't have to feel bad about scrapping a cap or two.

Frequency Loss

GHz	dB
2.0	−54
2.5	−43
3.0	−31
3.42	−4.4
3.45	−1.1
3.48	−4.0
4.0	−20
4.5	−22
5.0	−27

Temperature Compensation or Crystals

By Dave Mascaro, WA3JUF

(From *QEX*, April 1996)

Since my "Hints & Bits" article in June 1994 *Cheesebits*, several homebrewers have asked questions about ovens and thermistors. Which unit is easiest to use? How are they wired into an existing unit? Does something as simple as a thermistor really work?

Why is temperature compensation a concern? The frequency of your transverter's LO can drift many kilohertz when you go roving or during seasonal temperature changes in your basement. Few amateurs think about the temperature drift of their transverters or commercially made transceivers when they take the equipment out to the field. I measured a 75-kHz shift on the nonovenized LO of a 10-GHz transverter. Reducing (if not eliminating) the frequency variable when working a weak station is very significant.

The easiest, cheapest and fastest way to temperature compensate a crystal is to use a PTC thermistor. The leaded thermistor KC004P is $2.31 from Digi-Key (800-344-4539) and can be added to any crystal circuit in 10 minutes. The unit is connected directly to 12 V dc. The Yaesu G9090019 thermistor comes with its own holder that slides over an HC-25/U crystal. The number for Yaesu USA is 310-404-2700.

First, quickly unsolder one lead of the thermistor without damaging the metalization. Solder tin the side of the crystal. Solder the flat side of the thermistor to the crystal case.

Solder the remaining lead to the +12-V line. Solder a small gauge wire from the case of the crystal to dc ground to complete the dc path. The thermistor is nominally 50 Ω at 25° C, so it draws several hundred milliamps for a short while, then settles down to an idling current of less than 30 mA when it reaches its operating temperature.

I found the 60° C unit to be hot enough, even for the temperature excursions my equipment sees in the attic. Even in winter the crystal temperature stabilizes after about five minutes, instead of drifting for hours. Adding a small styrofoam or insulated cover over the crystal will further stabilize the frequency.

Don't get hung up on netting crystals to an exact frequency. Adding a heater will age the crystal in addition to moving its frequency, so allow the transverter to stay on for several days to complete the aging process. You may not be able to pull it back to the original frequency with the crystal trimmer. You can either make a note of the exact frequency and use it that way, or figure out how far the heated crystal moved and order another crystal based on that frequency delta and heater temperature.

I have added thermistor compensation to several SSB Electronic transverters, which all adjusted back to the original frequency. The thermistors work great on reference crystal oscillators for PLL sources. Adding the $2.31 thermistor to a DEM no-tune transverter produces a rock-solid LO that won't budge even in the hot sun.

Mixers, Etc, for 5760 MHz

A modular building block transverter

5760 MHz may be the least utilized amateur microwave band. Few construction articles have been published and very little equipment is available. Yet there is a wealth of surplus components available which are usable on this band as building blocks for a transceiver. The missing pieces, readily constructed by the microwave experimenter, are described and integrated into a system.

By Paul Wade, N1BWT

(From *Proceedings of 1992 Microwave Update*)

Introduction

5760 MHz may be the least utilized amateur microwave band. Few construction articles have been published and very little equipment is available. Yet there is a wealth of surplus components available which are usable on this band as building blocks for a transverter. The missing pieces are readily constructed by the microwave experimenter, so that more stations could get on the air. In the week after my transceiver became operational, I contacted five other stations; For four of us, it was our first contact on this band.

The wavelength for this band is 6 cm, so a dish antenna will have only 3 dB less gain than on 10 GHz—large gains are available from modest size antennas. On the other hand, amplifiers capable of providing several watts of power are much more readily available for 5760 than for 10 GHz. Combining them gives a modest station a really impressive ERP, so DX possibilities are good.

Mixer

The heart of any transverter system is the mixer, and there are few choices available for 5760 MHz. Unlike other components, surplus mixers for this band are scarce, so homebrewing is necessary. One option, the KK7B no-tune transverter[1] for this band has a simple bilateral mixer, used for both receive and transmit, so that switching is needed to utilize separate power amplifiers and receive preamps. Having separate mixers for transmit and receive is preferable so that each path may be optimized for its function.

[1] Notes appear at the end of this section.

The KK7B transverter has a 1296 MHz IF, probably because of the difficulty of making sharp filters, or any other high-Q circuit, reproducibly on a printed-circuit board at this frequency. The dimensions are too small and critical for normal printed-circuit tolerances.

Another transverter, with separate transmit and receive mixers, was described[2] (in German) by DJ6EP and DC0DA, and subsequently reprinted[3] in *Feedpoint* and *73*. The latter also described a modification to use a surplus Phase-Locked Microwave Source as the local oscillator, and made PC boards available, making it even more attractive. A unit was assembled, with abysmal results. There was no apparent mixing taking place and the only output was strong LO leakage. Closer examination of the mixer circuit suggested that it might be a harmonic mixer, operating with a half-frequency LO. This was confirmed when we located someone who could fake enough German to translate the article. At 5.6 GHz, the LO input impedance is effectively a short circuit, and measures exactly that.

It was obviously time for a new design. Some time ago, I designed and built a series of balanced mixers[4,5] using 90° hybrid couplers[6] from 1296 to 5760 MHz. Since these had worked well as receivers, two mixers were integrated with a third 90° hybrid coupler as a power splitter on a small Teflon PC board. The layout is shown in Figure 1. As expected, it worked well as a receive mixer, with about 7 dB of conversion loss. However, it worked poorly as a transmit mixer, with transmit conversion loss of around 25 dB.

This nonreciprocal performance was a mystery until

Rick, KK7B, steered me to an article explaining why a 90° hybrid-coupler works as a downconverter but not as an upconverter. I had only worked out the downconverter case and assumed that it would be reciprocal.

One reason for choosing the 90° hybrid coupler is that it is a low-Q structure and uses wide, low impedance transmission lines, so that dimensions are not extremely critical and performance should be reproducible.

The KK7B mixer[1] used a 6/4l "rat-race" coupler,[6] so the next version, shown in the photograph of Figure 2 used this structure for the transmit mixer. Line widths are somewhat narrower than the 90° hybrid coupler, but it is still a low-Q structure, so it should still be reproducible. This unit had much better transmit performance, about 8 dB of conversion loss, but its noise figure was not quite as good as the original receive mixer, so the original receive mixer was retained.

The final version integrates the "pipe-cap" filters like those in the DJ6EP[2] transverter onto the mixer board. These are copper plumbing pipe caps for three-quarter inch copper tubing, with probes 7/32 inch long and tuned with an 8-32 screw; dimensions are from the measurements made on individual filters. PC board layout is shown in Figure 3, and the only other components on the board are the mixer diode pairs and a 51 ohm chip resistor termination. IF attenuators like those in some of the no-tune transverters would also fit, and are recommended for the transmit side. No through holes are needed for grounding—the radial transmission line stub acts as a broadband RF short. The diodes I used (Hewlett-Packard HSMS-8202) are inexpensive Ku-band mixer diode pairs; they and the mixer boards are available from Down-East Microwave.

Local Oscillator

Microwave local oscillators normally start with a crystal in the 100 MHz range, followed by a string of multipliers. For 5760 MHz, a multiplication factor of 50 to 60 is necessary—not an easy task. Fortunately, there are many surplus Phase-Locked Microwave Sources (often called PLO bricks) available, made by companies such as Frequency West and California Microwave. These units were used in the 5.9-6.4 GHz communications band, and provide more than enough LO power for the mixer (a 6 dB attenuator was needed with mine). Some units have an internal crystal oven; after a few minutes warm-up, stability is comparable to a VHF transceiver. Operation and tune-up of these units has been described by KØKE,[8] WD4MBK,[9] and AA5C.[10] The sources can be used unmodified to provide high-side LO injection, above 5760 MHz, or modified[11] to operate below 5760 for normal low-side injection. Unless you are obsessive about direct digital readout, high-side injection, using LSB and reverse tuning, is perfectly acceptable. For CW operation, there is no difference.

Most of the available sources operate on –20 volts. This is only a problem for portable operation. WB6IGP has described[12] a +12 volt to –24 volt converter, and surplus potted converters are occasionally found. A three-terminal regulator IC provides the –20 volts. In order to prevent switching noise generated by the converter from reaching the LO, this is all contained in a metal box with RFI filtering on both input and output.

Filter

A good filter is essential for a serious microwave station, particularly for mountaintop operation. Most accessible high places are crowded with RF and microwave sources, so the RF environment is severe. I've seen unfiltered no-tune transverters for other bands fold up and quit in mountaintop environments.

The best filter I've tried is a waveguide post filter, such as the 10 GHz ones described by N6GN.[13] It is easily built using only a drill, tuning is smooth and non-critical, and the performance is excellent. Glenn was kind enough to calculate dimensions for 5760 using standard waveguide and hobby brass tubing, as shown in Figure 4. Dimensions are for WR-137 waveguide for 5800 MHz; reducing the spacings a hair will give a little more tuning range. I built two units; the second, with careful fit and flux cleaning had 0.4 dB of loss, while the first, with sloppy fit and soldering, had 0.5 dB of loss. Both units measured as shown in Figure 6, with steep skirts (135 MHz wide at 30 dB down) and no spurious responses detectable (>70 dB down). Tuning was smooth and easy; with high-side LO injection, the LO and image frequencies are out of the tuning range, so 5760 is the only output that can be found while tuning.

Construction Hint:

Make sure the holes are carefully measured and centered in the waveguide. Centerpunch lightly. Using a drill press, start the holes using a center drill, then drill them out a few drill sizes undersize. Then enlarge them one drill size at a time until the tubing is a snug fit. Solder on a hotplate using paste rosin flux.

Power amplifier

Semiconductor devices capable of good power and gain at this frequency are not as readily available as complete surplus amplifiers. Both TWT (Traveling Wave Tube) and solid-state amplifiers in the one to ten watt power range are fairly common. TWT operation was recently described by KD5RO.[14] These units usually have 30+ dB of gain and typically require one milliwatt (0 dBm) drive power.

Since our mixer has a maximum linear output of around –7 dBm, a small intermediate amplifier is needed. A GaAsFET stage could be used, but a broadband MMIC amplifier is also usable — some of the common MMICs have usable gain left at this frequency. WØPW[15] has described several amplifiers using MAR-8 devices. After seeing several unpublished papers about cures for oscillating MAR8s in 3456 MHz no-tune transverters, I chose to use the MSA-0986, which is unconditionally stable, at the cost of reduced low frequency gain. It has a flat gain of approximately 7 dB from about 0.1 to 4 GHz, rolling off to zero at about 8 GHz. At 5760, the two-stage amplifier shown in Figure 6 has 8.3 dB of gain; the 1/8 inch disc on the input line is adjusted to peak the output at 5760, yielding an additional dB, then soldered in place. Gain is still broadband, about 14 dB over the 903 to 3456 MHz bands, falling to zero at 8 GHz.

Construction is with minimal lead length on a scrap of Teflon PC board, with a soldered sheet brass enclosure. Keep the enclosure dimensions smaller than WR-137 waveguide cross-section and it won't oscillate. The blocking capacitors marked with an asterisk should be high quality chip capacitors.

Preamplifier

The WB5LUA GaAsFET preamps[16] are completely satisfactory—they work. My attempt to design a single stage amplifier was noteworthy only because it oscillates within 50 MHz of 5760.

Switching

KH6CP/1 has described two[17,18] transverter switching units that are a good starting point. Since most surplus coax relays operate on 28 volts, I tap off some negative voltage from the PLO supply through a 7915 three-terminal regulator IC to pro-

vide –15 volts to one side of the relay. The other side is connected to the +12 volt input to activate it with a total of ≈28 volts.

Surplus

There is a wealth of surplus components available that were originally used in the 5.9-6.4 GHz communications band. Most of these are broad enough to cover 5760 as well. As many commercial systems are upgraded, some of it appears dirt-cheap at hamfests—I've seen scrap metal dealers with truck-loads.

Particularly useful components are circulators and isolators,[9] which allow RF to flow in only one direction, protecting amplifiers from load mismatch (and often prevent oscillation).

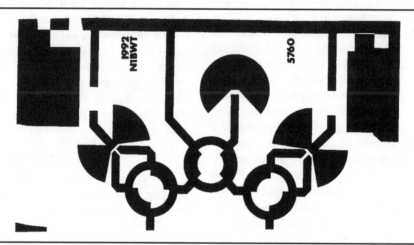

Fig 1—First dual mixer layout.

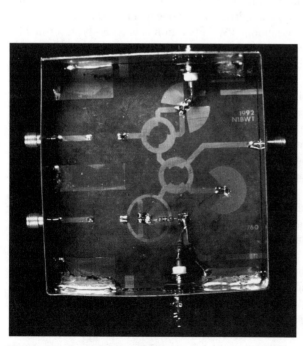

Fig 2—Revised dual mixer.

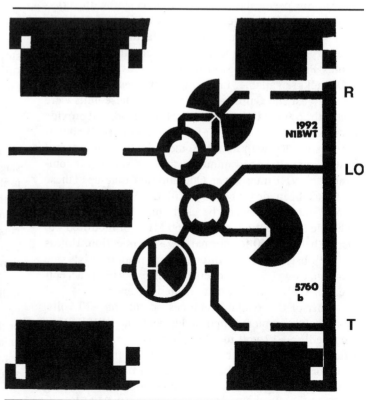

Fig 3—Layout of dual mixer with filters.

I use isolators between the LO and mixer, between the preamp and filter, and, of course, between the power amp and antenna. Other useful components are waveguide sections, useful for making filters, horn antennas, directional couplers, attenuators, and diode detectors, both coax and waveguide. Some filters can be retuned to 5760; I found a waveguide filter for 6.2 GHz that just barely reached 5760 at the tuning limit, but it had more loss than the homemade ones. Slotted lines are great for measuring VSWR and impedance. Even the cabinet for my transverter originally held a now-defunct TWT amplifier.

Block Diagram

Fig 7 is a block diagram of the current state of my transverter. Of course, with the building block approach, any of the blocks can be changed to improve the system.

Weak Signal Source

On a lightly populated band, having a weak signal source available greatly improves your confidence. I have been using an 1152 MHz signal source[20] based on the no-tune LO scheme, with a waveguide diode detector acting as a multiplier to 5760 MHz (or 10368 MHz). Recently I put together a much simpler version, based on an 80 MHz crystal oscillator of the type used in computers. Multiplying by 72 equals 5760 MHz, with no other harmonics near microwave calling frequencies. As shown in Fig 8, the oscillator drives an MMIC. K1TR gave me some MWA320 MMICs, which run fine on 5 volts like the oscillator, and I threw together a prototype on a piece of perfboard.

The oscillator output is a 5 volt square wave, so it is seriously overdriving the MMIC. As you would expect, the output waveform looks terrible, probably full of harmonics! Connect-

ing the output to a piece of waveguide (which acts as a high-pass filter) gives an S9 signal across the basement at 5760.150 MHz.

Higher frequency MMICs, like the MAR or MSA series, and better microwave construction would probably work even better.

Conclusion

The transverter system I have described started out as a mixer, then other modular building blocks were added to make a system. Many of the blocks are from surplus sources, others homebrewed as needed. In its current state, it's a good portable system for mountaintopping, but not capable of moonbounce. However, the building block approach can be used to assemble a system at any level of performance without limiting future enhancements.

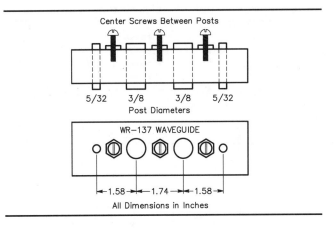

Fig 4—Waveguide filter for 5760 MHz.

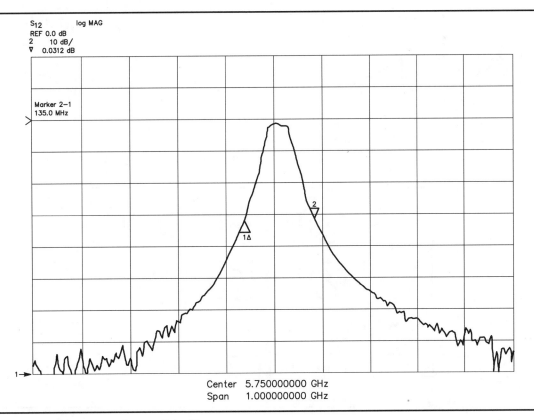

Fig 5—Measured performance of waveguide filter.

Bibliography

[1]R. Campbell, KK7B, "A Single-Board Bilateral 5760-MHz Transverter," *QST*, October 1990, pp 27-31.

[2]R. Wesolowski, DJ6EP, and J. Dahma, DC0DA, "Ein 6-cm-Transvertersystem moderner Konzeption," *cq-DL*, 1/88, pp 16-18.

[3]C. L. Houghton, WB6IGP, "Above and Beyond," *73*, December 1990, pp 61-62.

[4]P. Wade, WA2ZZF, "A High-Performance Balanced Mixer for 1296 MHz," *QST* September 1973, pp 15-17.

[5]P. Wade, WA2ZZF, "High-Performance Balanced Mixer for 2304 MHz," *ham radio*, October 1975, pp 58-62.

[6]H. S. Keen, W2CTK, "Microwave Hybrids and Couplers for Amateur Use," *ham radio*, July 1970, pp 67-61.

[7]K. Britain, WA6VJB, "Cheap Microwave Filters," *Proceedings of Microwave Update '88*, ARRL 1988, pp 159-163

[8]K. R. Ericson, K0KE, "Phased Lock Source Update," *Proceedings of Microwave Update '87*, ARRL, 1987, pp 93-95.

[9]C. Osborne, WD4MBK, "Surplus Microwave Local Oscillators, Evaluating and Modifying Them," *Proceedings of Microwave Update '88*, ARRL, 1988, pp 33-41.

[10]G. McIntire, AA5C, "Phase-Locked-Microwave Sources," *Proceedings of Microwave Update '91*, ARRL, 1991, pp 113-136.

[11]C. L. Houghton, WB6IGP, "Above and Beyond," *73*, November 1991, pp 66-68.

[12]C. L. Houghton, WB6IGP, "Above and Beyond," *73*, July 1990, pp 68-69.

[13]G. Elmore, N6GN, "A Simple and Effective Filter for the 10-GHz Band," *QEX*, July 1987, pp 3-6.

[14]D. Halliday, KD5RO/2, "TWTs and Klystrons: How They Work and How to Use Them," *Proceedings of Microwave Update '91*, ARRL, 1991, pp 234-246.

[15]Hilliard, D., W0PW, "2 GHz to 6 GHz Power Amplifiers," *Proceedings of Microwave Update '87*, ARRL, 1987, pp 78-92.

[16]A. Ward, WB5LUA, "Simple Low-Noise Microwave Preamplifiers," *QST*, May 1989, pp 31-36.

[17]Z. Lau, KH6CP/1, "T/R Switching Low-Power 903 and 1296 MHz Transverters," *QEX*, July 1992, pp 16-17.

[18]Z. Lau, KH6CP/1, "A VHF/UHF/Microwave Transverter IF Switch," *QEX*, August 1988, pp 34.

[19]K. Britain, WB5VJB, "Circulators and Isolators," *Proceedings of the 25th Conference of the Central States VHF Society*, ARRL, 1991, pp 31-32

[20]P. Wade, NIBWT, "Weak Signal Sources for the Microwave Bands," *Proceedings of the 18th Eastern VHF/UHF Conference*, ARRL, 1992, pp 91-93.

Additional reference: Chang, K.W., Chen, T.H., Wang, H. and Maas, S.A., "Frequency Upconversion Behavior of Singly Balanced Diode Mixers," *IEEE Antennas and Propagation Society Symposium 1991 Digest*, Vol 1, pp 222-225, IEEE, 1991.

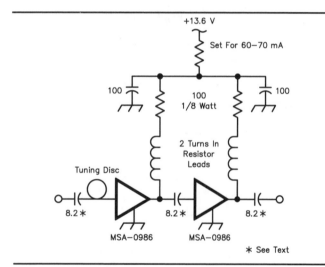

Fig 6—MMIC amplifier.

Fig 8—5760 MHz weak signal source.

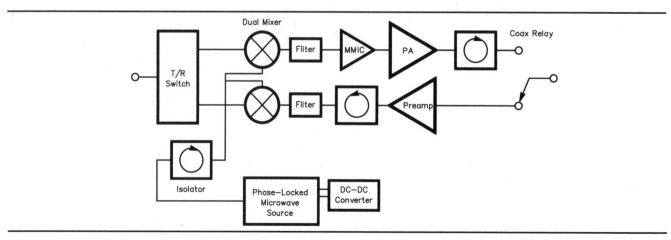

Fig 7—5760 MHz transverter block diagram.

Building Blocks for a 10 GHz Transverter

By Paul Wade, N1BWT

(From Proceedings of The 19th Eastern VHF/UHF Conference)

Introduction

This is not a description of a complete no-tune transverter for 10 GHz SSB and CW, but rather a series of building blocks that comprise the critical parts of a transverter. These will ease the development of a working narrowband system, but some skill and experience is required at 10 GHz. There are still enough vagaries at this frequency that a single-board transverter would probably not work as assembled—building blocks can be individually checked before assembly into a system.

One obstacle to the development of reproducible 10 GHz equipment is the lack of affordable devices to provide gain, so the first building block is a two-stage amplifier to provide adequate gain in the several places required to make a working system. Other blocks include a dual mixer, with separate mixers for transmit and receive, an LO multiplier to generate a 10 GHz local oscillator, and a power amplifier for modest power output, plus some blocks that others have previously described that don't need reinventing. All these blocks can be put together as a transverter, such as the block diagram in Figure 13 below.

Background

My previous 10 GHz narrowband system used a surplus waveguide balanced mixer, which performs reasonably well with only an LO, filter, and antenna as a low-power system. However, the low power limits range, and we have not made any contacts beyond the range of a good wideband FM Gunnplexer system. The next step, integrating the waveguide mixer into a high-performance system with transmit and receive amplifiers would require multiple switches and waveguide to coax transitions, so packaging it for portable operation would be unwieldy. A new system, using printed microstrip circuitry, was in order. At Microwave Update '92 in Rochester,[1] I presented a description of my modular building-block approach to assembling a transverter for 5760 MHz.

[1] Notes appear at the end of this section.

Since the 5760 transverter—designed with inexpensive, available parts—works well, I wondered whether a 10 GHz system could be made with the same approach.

As this project neared completion, another 10 GHz transverter, by KH6CP, appeared in *QST*. Zack's design[2] is also constructed as a series of building blocks, and someone contemplating building a 10 GHz system would do well to study both this version and Zack's, and perhaps choose the block that best fit available components and talents.

Two-Stage Amplifier

One obstacle to the development of reproducible 10 GHz equipment is the lack of affordable devices to provide gain. The no-tune transverters for the lower microwave bands use low-cost silicon MMICs for the gain elements, but these don't work at 10 GHz and gallium arsenide MMICs are not cheap. WB5LUA has described excellent low-noise preamps,[3] which are also useful as low-power amplifiers, but the GaAsFETs that Al used are rather pricey. Inexpensive GaAsFETs are available, so a reproducible design could be used as a gain block like an MMIC.

Amplifier Design

The design goal for this amplifier was simple: to utilize the cheapest readily available device that provides good performance at 10 GHz. The best I found is the Avantek ATF-13484 at $4 from Down East Microwave. A bit of computer calculation with S-parameters yielded a straightforward design that could be printed on common $1/32$-inch thick Teflon PC board. The circuit has a transmission line and a stub at the output, plus bias decoupling networks. A layout with two identical stages is shown in Figure 1.

A proper bias network is one key to disaster-free operation. I prefer the active bias circuit described by WB5LUA[3] for his low-noise GaAsFET preamps. The schematic diagram in Figure 8 shows the two-stage amplifier with active bias circuits. The component values shown result in an operating point of about 3.5 volts at 35 mA for each stage.

Amplifier Performance

Measured gain, shown in the curves of Figure 2, is about 15 dB at 10368 MHz. This performance was achieved after minor trimming of the stubs nearest to the connectors, probably to compensate for the connector mismatch. The one-dB compression point was around +13 dBm, a modest amount of power at 10 GHz. Although the amplifier was designed to work with the ATF-13484, other available small GaAsFETs also work fine. Units were assembled using the Mitsubishi MGF-1302 and MGF-1412, which both provided more gain than the original with no tuning, about 21 and 19 dB respectively. Curves for these amplifiers are also shown in Figure 2.

Noise figure measured around 5 dB at operating current, and would probably be lower at lower current with a bit of retuning. This is hardly a super front end, but it is respectable for the cost, and much better than the bare mixers many of us have been running.

Figure 2 also shows the predicted ATF-13484 performance for lossless components. Measured gain for the ATF-13484 unit is a bit lower, as might be expected with real components at this frequency.

Amplifier Construction

Construction is with minimal lead length on Teflon PC board, with a soldered sheet brass enclosure. Components marked with an asterisk are chip capacitors or resistors, and the blocking capacitors should be high quality chip capacitors. Parts placement can be seen in the photograph, Figure 7.

Achieving minimum lead length takes a bit of effort, but is worth it. I do it by bending the ground leads down right at the package, as illustrated in the sketch by DB6NT[4] shown in Figure 9. Then I drill a hole in the PC board, which just fits the package body, and file the sides just enough for the ground leads to slip through. Then I push the device down until the input and output leads start to bend up, so the ground leads are as short as possible. The ground leads are then bent over and soldered to the ground side of the board right at the edge of the hole.

Stability

The key to stability in these amplifiers is minimum ground lead inductance, achieved by keeping the leads as short as possible. In one of the units, I assembled MGF-1412 devices in smaller holes drilled for MGF-1302 (I dropped my last 1302 on the floor and couldn't find it). This resulted in slightly longer ground leads, perhaps 0.5 mm, and an extremely unstable amplifier, which couldn't be tamed.

Computer simulation with additional ground inductance confirmed this behavior, which was only cured by getting more MGF-1302s and installing them. Computer analysis with minimum ground lead inductance predicts sta-

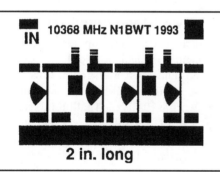

Fig 1

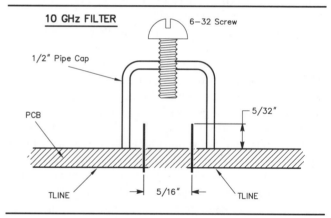

Fig 3

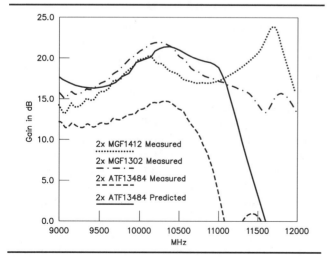

Fig 2

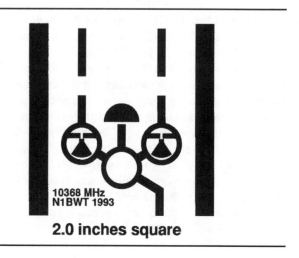

Fig 4

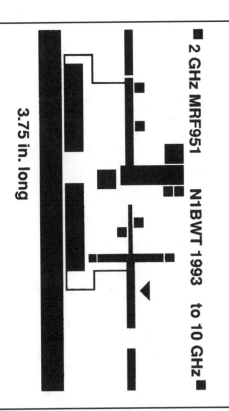

Fig 5–PC layout for 10 GHz LO multiplier.

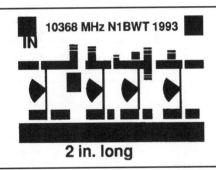

Fig 6–PC layout for 10 GHz power amplifier.

Fig 7—Photograph of two-stage GaAsFET amplifier for 10 GHz.

bility above 4 GHz, but shows a potential instability at lower frequencies, which is readily cured by adding resistance from gate to ground. A 220 ohm chip resistor was adequate for stability and does not appear to affect the 10 GHz gain; the only drawback is that it draws current from the negative bias supply, so that some of the smaller voltage inverter circuits may not provide enough current.

Dual Mixer

The 5760 transverter uses a dual mixer, with separate mixers for transmit and receive, so that each path may be optimized for its function, and separate amplifiers for transmit and receive are easily added. In fact, for 5760 I used different types of mixers for the two paths. However, when I tried to scale this design to 10 GHz, the 90° hybrid coupler balanced mixer used for the receive was no longer practical—the wide, low impedance transmission lines would be wider than their length. One alternative would be the use of thinner Teflon PC board material, which would make the lines proportionally narrower; however, this would make the high impedance line sections very narrow. Since I wanted to use the more common $1/32$ inch thick material, the other alternative was to change the receive mixer design to the $6/4\,1\lambda$ "rat-race" coupler type used in the transmit mixer. Now the linewidths are reasonable for my basement PC board techniques.

The complete mixer integrates the "pipe-cap" filters onto the mixer board, as originally described[5] (in German) by DJ6EP and DCØDA, in a transverter for 5760 MHz. The filters are made from copper plumbing pipe caps for $1/2$-inch copper tubing, tuned with a 6/32 brass screw, with probes of #24 wire projecting $5/32$-inch above the ground plane inside the filter. The probes are $5/16$-inch apart A cross-section sketch of a pipe-cap filter is shown in Figure 3. I started with dimensions from the measurements WA5VJB[6] made on individual filters. However, the difference in construction techniques seems to make a significant difference in performance at 10 GHz, so I chose the length after making the measurements plotted in Figure 10, for a 3 dB bandwidth of about 140 MHz with loss of about 4 dB.

PC board layout is shown in Figure 4. Other than the filters, the only components on the board are the mixer diode pairs and a 51 ohm chip resistor termination. IF attenuators like those in some of the no-tune transverters would also fit, and are recommended for the transmit side. No through holes are needed for grounding—the radial transmission line stub acts as a broadband RF short. The diodes I used (Hewlett-Packard HSMS-8202) are inexpensive Ku-band mixer diode pairs.

Mixer Performance

Performance of the 10 GHz dual mixer is not as good as the 5760 version—we are pushing the limits of these simple techniques. Conversion loss is around 13 dB, partly due to the 4 dB filter loss. This is part of the compromise for making it simple and inexpensive. However, we can overcome the loss with the amplifier described above.

Mixer performance is dependent on the local oscillator

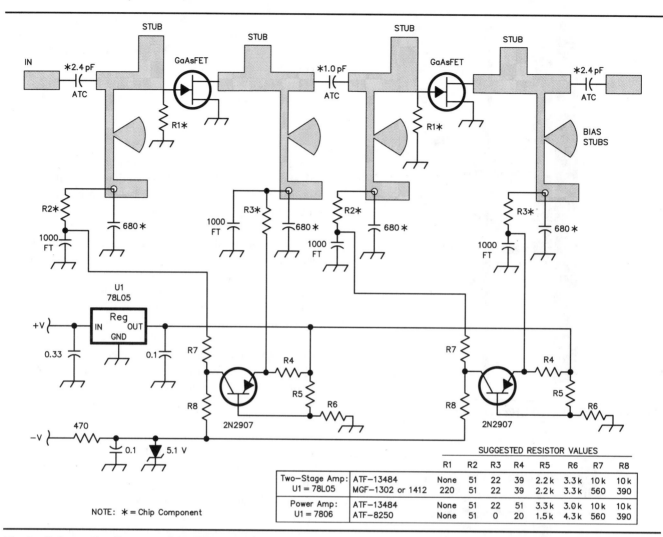

		R1	R2	R3	R4	R5	R6	R7	R8
Two–Stage Amp: U1 = 78L05	ATF–13484	None	51	22	39	2.2 k	3.3 k	10 k	10 k
	MGF–1302 or 1412	220	51	22	39	2.2 k	3.3 k	560	390
Power Amp: U1 = 7806	ATF–13484	None	51	22	51	3.3 k	3.0 k	10 k	10 k
	ATF–8250	None	51	0	20	1.5 k	4.3 k	560	390

SUGGESTED RESISTOR VALUES

NOTE: ✻ = Chip Component

Fig 8—Schematic diagram of 10 GHz amplifier.

power. The HSMS-8202 diodes require less LO power than most, which is usually an advantage. Low LO power will limit the maximum output power from the transmit mixer, while excess LO power will increase the receive mixer noise figure, so the optimum point is a compromise. Since the LO power splitter on the mixer board is not perfect, one mixer may get more LO than the other, so test both sides as both transmit and receive.

Mixer Construction

Construction is with minimal lead length on a Teflon PC board, with soldered sheet brass around the perimeter for SMA connector attachment. This is the procedure I use: The copper pipe-cap filter should be installed first, on the ground-plane side of the board. In preparation, I drill tight-fitting holes for the probes and make clearance holes in the ground plane around the probe holes. Then I measure from the holes and scribe a square on the ground plane that the pipe cap just fits inside. Next I prepare each pipe cap by drilling and tapping (use lots of oil) the hole for a tuning screw, then flattening the open end by sanding on a flat surface. Then I apply resin paste flux lightly to the open end and the area around the screw hole. A brass nut, added to extend the thread length, is held in place by

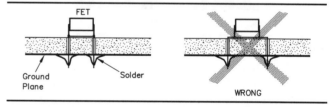

Fig 9—Achieving minimum lead length (by DB6NT).

a temporary stainless-steel screw (Solder won't stick to it). Then I center the open end in the scribed square on the PC board—the flux holds it in place. Finally, I fit a circle of thin wire solder around the base of each pipe cap and nut, push down gently, and heat each pipe cap for few seconds with a propane torch until the solder melts and flows into the joints.

After everything cools, the temporary stainless-steel screw should be replaced with $^3/_4$- inch long brass tuning screws and locknuts. The remainder of the assembly is performed with a soldering iron, using the photograph of Figure 11 as a guide.

Local Oscillator

Microwave local oscillators normally start with a crystal

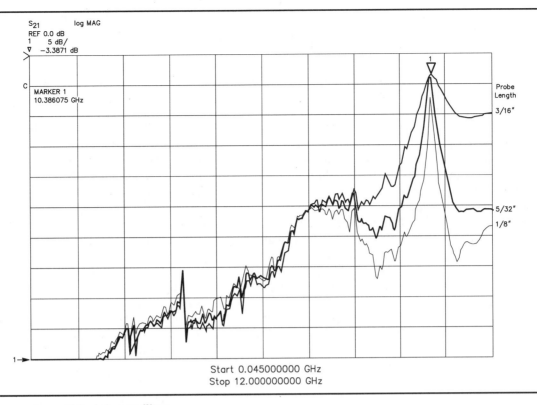

C

MARKER 1
10.386075 GHz

Probe
Length

3/16"

5/32"
1/8"

1→

Start 0.045000000 GHz
Stop 12.000000000 GHz

Fig 10—Measurements on ¹/₂" pipe-cap filter.

in the 100 MHz range, followed by a string of multipliers. For 10 GHz, a multiplication factor of 100 or more is necessary—not an easy task. However, most of the work has been already done in making local oscillators for the lower microwave bands, so only one more stage of multiplication is needed. I will describe this approach as well as the surplus alternative.

Phase-Locked Microwave Sources

If you are fortunate enough to find them, there are many surplus Phase-Locked Microwave Sources (often called PLO bricks) available, made by companies such as Frequency West and California Microwave. Most of the available surplus units were used in the 11-12 GHz band, and provide adequate enough LO power for the mixer. Some units have an internal crystal oven; after a few minutes warm-up, stability is comparable to a VHF transceiver. Operation and tune-up of these units has been described by KØKE,[7] WD4MBK,[8] and AA5C.[9] The sources can be used unmodified to provide high-side LO injection, above 10368 MHz, or modified[10] to operate below 10 GHz for normal low-side injection. Unless you are obsessive about direct digital readout, high-side injection, using LSB and reverse tuning, is perfectly acceptable. For CW operation, there is no difference.

Most of the available PLO sources operate on −20 volts. This is only a problem for portable operation. WB6IGP has described[11] a +12 volt to −24 volt converter, and surplus potted converters are occasionally found. A three-terminal regulator IC provides the −20 volts. In order to prevent switching noise generated by the converter from reaching the LO, all this should be contained in a metal box

Fig 11—Photograph of 10 GHz dual mixer.

with RFI filtering on both input and output.

LO Multiplier to 10 GHz

The PLO bricks are relatively scarce, and they do require a substantial amount of power, so a multiplication from a lower frequency is an alternative. One attractive choice is the 2160 MHz LO for 2304 MHz by KK7B; 2160 multiplied by five equals 10800 MHz, which is perfect for a 432 MHz IF at 10368 MHz. Since the pipe-cap filters have a bandwidth of about 140 MHz, using 432 as the IF will provide much more LO rejection.

I've never had much success with GaAsFET frequency

multipliers, so a diode multiplier seemed like the way to go. A recent article in *Microwave Journal* described using back-to-back diodes to favor odd harmonics in millimeter-wave frequency multipliers. Why not at lower frequencies? I used the same diode pair (HSMS 8202) used in the mixer, with both diodes connected from a 50 ohm line to ground to form a back-to-back pair. The diodes are installed by cutting a small slot, slightly larger than the diode package, through the board right next to the transmission line. The diode leads are bent so that the single common lead is soldered to the transmission line and the pair of leads on the opposite side of the package are bent and soldered to the ground side of the board.

The diodes are followed by a pipe-cap filter, just like the ones on the mixer board. With approximately +20 dBm applied at 2160 MHz, output was about –5 dBm at 10800 MHz; our two-stage amplifier will bring this up to an adequate LO level. The multiplier output is relatively insensitive to the input power. To test the output spectrum, I turned the tuning screw in the pipe-cap filter; the only other response was at 6480 MHz, the third harmonic, so the back-to-back diodes do favor odd harmonics. Replacing them with a single diode reduced the output at 10800 MHz by about 6 dB, further proving the effectiveness of this configuration.

The KK7B[12] 2160 MHz source has an output of about +10 dBm, so some amplification is required to drive the multiplier. W3HQT suggested the MRF-951 bipolar transistor for this stage, and it works well, providing more than +20 dBm output with about 9 dB of gain, operating at 8 volts and 50 mA. PC board layout, shown in Figure 5, includes the pipe-cap filter; the multiplier diode is opposite the printed arrow. The schematic diagram is shown in Figure 12.

Construction is like the two-stage amplifier, except that everything is much larger! For tune-up, I assemble everything but the diodes and pipe-cap filter probes, then bridge the filter with a brass strap, and test the amplifier at 2160 MHz (The amplifier alone makes a nice power amplifier for a 2304 MHz no-tune transverter). After verifying that the amplifier works, I remove the strap and install the diodes and probes. Then the filter is tuned for output power at 10800 MHz (the first response as the screw is turned inward) and the amplifier tweaked for maximum output.

Power Amplifier

The two-stage amplifier described above has a one-dB compression point around +13 dBm, or 20 milliwatts. This is hardly QRP at 10 GHz, since we have been using barefoot mixers with 20 dB less output, but more power is always attractive. I reviewed the available higher-power devices, and most of them seemed to have little gain left at 10 GHz, so many stages would be required. One possibility was the AT-8250 (ATF-25170), offering moderate power with usable gain. A two-stage amplifier was designed with a ATF-13484 as a driver for increased gain, with the layout shown in Figure 6.

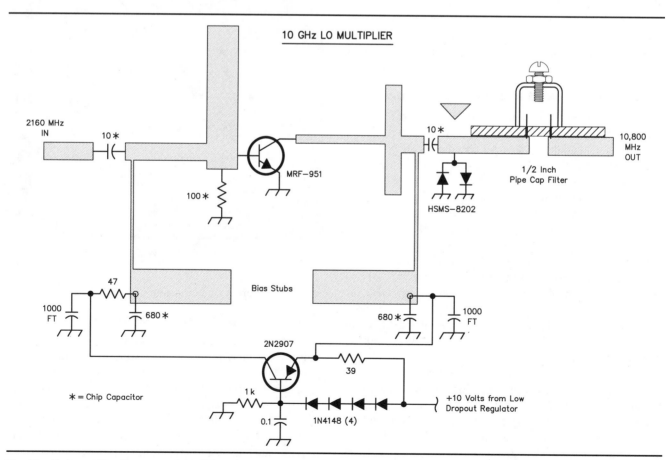

Fig 12—Schematic diagram of 10 GHz LO multiplier.

The amplifier has about 13 dB of gain and a one-dB compression point of about +19 dBm; a bit more drive yields 100 mW output—respectable power at 10 GHz.

To provide the additional power, the AT-8250 is biased to 5 volts and 50 mA. by changing component values in the bias network, as shown in Figure 8. Otherwise, it is so similar to the two-stage amplifier that a separate schematic is unnecessary.

Tune-Up

This is not a no-tune design. Components and PC boards are just not consistent enough for that. However, they are close enough that things will work if assembled correctly, but need some tuning for optimum performance. Obviously, the tuning screws in the pipe-cap filters must be tuned to frequency, but the amplifiers and mixers also benefit from fine tuning. This is accomplished by sliding little pieces of thin copper or brass around on the printed transmission lines, looking for "hot-spots" which significantly change the output. If the output increases, the metal piece can be soldered in place to make it permanent. On the other hand, if the output decreases, a bit of the printed metal may be trimmed off with a sharp knife. Make small changes! It doesn't take much at 10 GHz, and a few small improvements can add up.

Transverter Block Diagram

Figure 13 illustrates one possible block diagram for a transverter using the building blocks described above. Expected signal levels are shown at each point. The isolators[13] shown are optional, but recommended to ensure stable operation if you have them—they frequently appear at our local flea markets at very reasonable prices. An isolator will allow RF to flow in only one direction, protecting amplifiers from load mismatch (and often prevent oscillation). For the preamp, I'd recommend the WB5LUA[3] design.

An output filter is highly recommended, particularly if mountaintop portable operation is contemplated. Most accessible high places are teeming with RF microwaves, so a good filter will keep them out of your transverter. One excellent one is the waveguide post filter described by N6GN,[14] which is easily built in a section of surplus waveguide and provides performance far superior to the pipe-cap filters. Since most 10 GHz antennas are fed with waveguide anyway, the filter fits well at the output.

Conclusion

Not surprisingly, working at 10 GHz is more difficult than the lower microwave bands. Everything is more critical—some things can be tuned by tightening an SMA connector (the only kind that works). Test equipment is necessary; you don't have to have a fancy lab, but having some way of measuring power and frequency is essential. All the equipment I used for tune-up has been obsolete for years, and most was found at flea markets. Later I was able to use a fancy Network Analyzer to generate some of the plots in this paper.

What was encouraging is that a circuit design can be analyzed on a computer, fabricated using basement printed-circuit techniques, built with tinsnips and soldering iron, and work rather reproducibly if built with care. On the other hand, the amplifier instability described above, which was caused by a little extra ground lead length, was a real eye-opener.

The building blocks described should make it possible for someone wishing to get on 10 GHz to assemble a transverter by building one piece at a time, checking it out, making contacts, adding on for increased performance, and learning more with each step. Also, all PC boards and critical parts are available from Down East Microwave, eliminating the often daunting task of finding that one last elusive part.

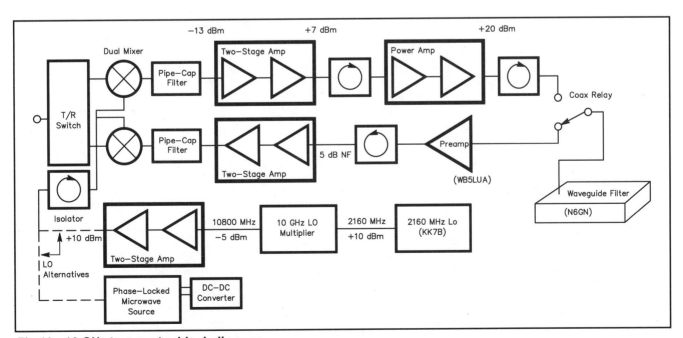

Fig 13—10 GHz transverter block diagram

Bibliography

[1]Wade, P., N1BWT, "Mixers, etc., for 5760 MHz," *Proceedings of Microwave Update '92*, ARRL, 1992, pp 71-79

[2]Lau, Z., KH6CP, "Home-Brewing a 10-GHz SSB/CW Transverter," *QST*, ARRL, May 1993, pp 21-28, and June 1993, pp 29-31.

[3]A. Ward, WB5LUA, "Simple Low-Noise Microwave Preamplifiers," *QST*, May 1989, pp. 31-36.

[4]M. Kuhne, DB6NT, "12 GHz LO for 24 & 47 GHz," *DUBUS Technik III*, DUBUS Verlag, 1992, pp 149-157

[5]R. Wesolowski, DJ6EP, and J. Dahms, DC0DA, "Ein 6-cm-Transvertsystem moderner Konzeption," *cq-DL*, 1/88, pp 16-18.

[6]K. Britain, WA5VJB, "Cheap Microwave Filters," *Proceedings of Microwave Update '88*, ARRL, 1988, pp 159-163.

[7]K. R. Ericson, K0KE, "Phase Lock Source Update," *Proceedings of Microwave Update '87*, ARRL, 1987, pp 93-95.

[8]C. Osborne, WD4MBK, "Surplus Microwave Local Oscillators, Evaluating and Modifying Them," *Proceedings of Microwave Update '88*, ARRL, 1988, pp 33-41.

[9]G. McIntire, AA5C, "Phase-Locked Microwave Sources," *Proceedings of Microwave Update '91*, ARRL, 1991, pp 113-136.

[10]C. L. Houghton, WB6IGP, "Above and Beyond," *73*, November 1991, pp 66-68.

[11]C. L. Houghton, WB6IGP, "Above and Beyond," 73, July 1990, pp 68-69.

[12]R. L. Campbell, KK7B, "A Clean, Low-Cost Microwave Local Oscillator," *QST*, July 1989, pp 15-21.

[13]K. Britain, WB5VJB, "Circulators and Isolators," *Proceedings of the 25th Conference of the Central States VHF Society,* ARRL, 1991, pp 31-32

[14]G. Elmore, N6GN, "A Simple and Effective Filter for the 10-GHz Band," *QEX*, July 1987, pp 3-5.

Just About as Cheap as You Can Get 10 GHz

The guts from a RADAR detector, a Ramsey Electronics FR10, and a 2N2222 modulator puts you on the air.

By Kent Britain, WA5VJB

(From *Proceedings Of 1996 Microwave Update*)

The Ramsey FR10 is easily converted into a very sensitive 30 MHz wideband FM microwave IF. The FR-10 has two filters in the IF section, simply bypassing the second filter and letting the first 10.7 MHz filter do its job, makes the FR10 a 200 kHz wide receiver. I chose to reduce the tuning range of the receiver to about .5 MHz, but this was a personal choice. The NE-602 Mix/Osc IC can be crystal controlled. See the February '92 issue of *73 Magazine*, page 21 for some suggested circuits:

Figure 1 is the only mod you have to do; replacing FL2 with a .01 mfd cap makes the FR10 a wideband receiver. The .01 goes from pin 3 to pin 5 of U2.

Figure 2 is the audio stage. The FR10 is designed to drive a speaker and is far too hot for headphones. (On Gunnplexer systems you need headphones to prevent feedback.) Omitting C35 (it was optional anyway!) and adding the 330 ohm resistor worked well.

Figure 3 was my tuning range mod. Replacing C12 with a 12-15 pF and adding another 12-15 pF across L3 tightened up the tuning range. Replacing C12 with a 22 pF will also work giving a 1 MHz tuning range and the ultimate would be the *73 Magazine* mod using a 40.7 MHz xtal.

Alignment is quick. Put the tuning control in the middle of its range and listen to a 30.0 MHz signal while adjusting L3.

Figure 1 also has a mod suggested by Al Ward. The 3.9K resistor across the quadrature detector coil broadens the FM detector response and improves audio response with highly deviated signals.

The RF head in Figure 4 is the guts from a RADAR Detector. To retune the Gunn oscillator from 11.5 to 10.25 GHz you'll need either a Wavemeter, Spectrum Ana-

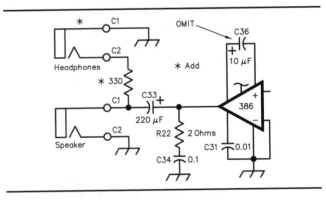

Fig 2—Ramsey FR-10 mods.

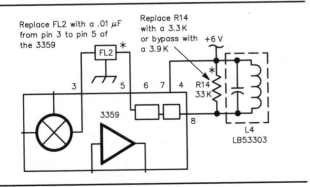

Fig 1—Ramsey FR-10 mods.

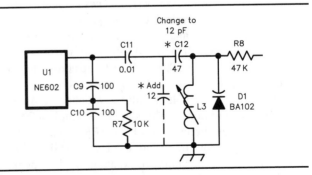

Fig 3—Ramsey FR-10 mods.

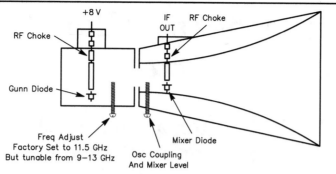
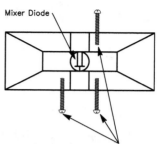

Notes:

The Osc and Mixer Level adjustment can be set for Max Power out or best Receive Sensitive. Normal setting is best rec.

Power output: 3–5 Milliwatts adjusted for best rec.
10–20 Milliwatts tuned for best TX
10–30 Milliwatts when used as a Beacon or Source with the Mixer Diode removed.

Normal IF out is 800 to 1200 MHz, but 30 MHz and even dc work just fine. (dc is for Doppler radar versions)

The rigid waveguide used in the mixer section and the horn antenna is very broad, and can be used from 9 to 33 GHz.

Waveguide Tuning Screws
These screws tune the mixer for best operation on 10.525 and 24.1 GHz.

For 10.3 GHz transceiver service remove all the screws.

But if you have 10 GHz test equipment, try putting them back in, and adjusting for best performance

Fig 4—Typical radar detector RF section.

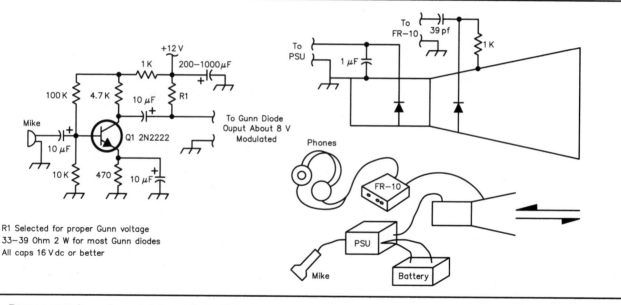

R1 Selected for proper Gunn voltage
33–39 Ohm 2 W for most Gunn diodes
All caps 16 V dc or better

Fig 5—Power supply and modulator connections.

lyzer, or an EIP type microwave frequency counter. A friend with test gear is very valuable at this point.

The 1K resistor and 39 pF capacitor in Figure 5 are typical values. I usually use a 10-80 pF trimmer and tweak for best sensitivity. The 0.01 to 0.1 μF cap across the Gunn diode keeps down noise and VHF parasitics.

It only takes 0.1 volt of audio to wideband FM modulate a Gunn oscillator. This modulator is simply a 2N2222 audio amp taken directly out of the *ARRL Handbook*. No ALC or compressor circuits are used. On 10 GHz FM you hear your own signal coming back, audio weak—speak up, too loud — back off the mike.

The 39 ohm 2 watt resistor drops the 12 volts to 8 volts at the Gunn diode. This can be tweaked for best results and is a typical value. Varying the 12 volt source will give you a limited tuning range.

Use shielded wire between the modulator and Gunn diode to keep down hum and RF rectification.

More complicated and versatile PSUs (Power Supply Modulators) are described in *The RSGB Microwave Newsletter Technical Collection* and the *RSGB Microwave Handbooks*. This design was kept as simple as possible. PS: Works with 24 GHz Gunns as well.

See ya on 10 GHz!

A "Tri-Band Booster"—or "Improving 2, 3 and 5 GHz at WA1MBA"

By Tom Williams

(From *Microwave Update '97*)

Problems with SHF from the Home Station

As some of you know, I run a home station that is active on all UHF and SHF bands (24 GHz to come soon). After being on the air for several years, and upgrading all aspects of the station, I have been bothered by the limited DX on 2, 3 and 5 GHz. Sure, I can work the mountaintoppers out to 100 miles or more, but there are good stations that I can't work during contests or other times. Simply put, this station could do a lot better in transmit power and receive noise figure.

As you can see, in Figure 1, there is a lot of cable resulting in significant losses. My original intent with this station was to put the LNAs as close as possible to the feed. This configuration is quite acceptable on the "low bands," but at 2304 and up the loss between the top-of-the-tower box and the feed is too great. I have about 7 dB of loss ahead of the 5 gig LNA, 4 dB on 3 gig and about 3 dB on 2 gig. This results in unacceptable system noise figures of 4 to 8 dB, whereas a good terrestrial station should achieve under 1 dB on all these bands. On the transmit side, antenna power levels are 2 watts on 2 gigs, and about a hundred milliwatts on 3 and 5 gigs. I figured that about 10 watts on each band would be enough to work the DX that I am missing. The only way to fix all of this is to put power amps and LNAs as close as possible to the feed.

The Plan

The perfect system would have a wideband high power amp (2 to 6 GHz) and a wideband LNA. This way, two components would be needed, along with simple relays to boost the transmit and receive signals, and my problem would be solved with an elegant system. But elegance is not in the cards.

Unfortunately, high power amplifiers that cover that range are very tricky to build and many thousands of dollars to buy. There are TWTAs that would cover the range, but this would mean having high voltages and a sensitive and expensive system up top. If I owned two miniature 10 W TWTAs (one for a spare), I would probably use them. If I ever design a solid state 10 W amplifier that covers all these bands, I will certainly share it with the community.

On the receive side, wideband preamps do exist. LNAs that cover these frequencies can be built with noise figures below 1.5 dB, but are very expensive, and no designs that I know of have been made public. One can achieve about 2+ dB with a GaAs MIMIC. I wanted to achieve under 1 dB noise figure, preferably 0.8 dB at 5 GHz, and 0.4 dB at 2 GHz. Fortunately, single band designs and parts that achieve this performance level are readily available.

Given all the above considerations, I set out to build separate amplifiers and preamplifiers for all three bands, and came up with a concept for the booster, based on connecting them with seven-port coaxial relays and appropriate power switching. I also included output power monitoring and a bypass connection in case any device becomes inoperative (or suspect).

Table 1

Power Loss and Gain Needed to Achieve 10 W at the Feed

Band	Power in Shack	Power at Feed	Cable Loss	Gain Needed to Achieve 10 W
2304	41 dBm (10+ W)	34 dBm (2+ W)	7 dB	6 dB
3456	33 dBm (2 W)	18.5 dBm (65 mW)	14.5 dB	21.5 dB
5760	37 dBm (5 W)	20.5 dBm (0.1+ W)	16.5 dB	19.5 dB

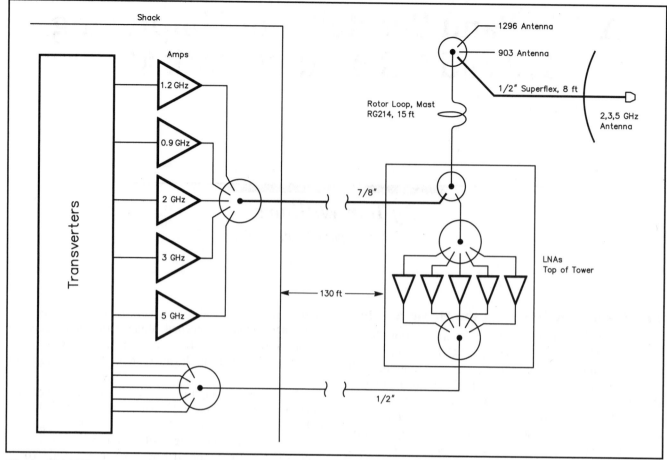

Fig 1—WA1MBA 0.9 through 5.7 GHz before booster.

The Experience

Decisions, decisions. . .

First, I had to decide whether to buy or build. It turns out that there are some sources of ready-made amplifiers for these bands, the higher power ones are very (very) expensive new, but there are occasional surplus amplifiers showing up at flea markets at a fraction of their original price. The flea market prices I've seen for 10 W 5 gig amps is $40 to $150. The Avantek ones take –28 Vdc and are very heavy. They are TWT replacements, having lots of low level gain so that they can be driven by as little as 0 dBm. Although excessive gain isn't a problem, I decided that the weight, and difficulty in making it a lot smaller would nix using this fine amplifier. The $150 five watt surplus unit I found is much smaller, and runs on +12 Vdc. Very nice. But I really wanted an even smaller package than this unit, and I wanted to build as much myself as possible, in order to gain experience. I put my five watter in service driving the transmit coax from the shack. I've also found 3 and 2 gig amps being sold from time to time, but they all were either too large or too expensive—and often both.

I decided to build a 2304 amp that WA3JUF (now W3KM) had described that uses a bipolar device normally designed for high power in class C operation. His design uses a heavy bias supply, resulting in class AB operation (clean signal) with about 10 dB of gain and an output around 10 W. This was just what I needed for this band. I asked Dave for some help, and he was glad to assist me with a spare device and a design. I found a dead amplifier (originally built for 903) that had a milled out slot for the transistor that was an exact fit. The resulting 2304 amplifier is quite small. The only disadvantage is that it needs heavy current at +28 V and –5 V.

For 3 GHz, I went with a DB6NT design that Eisch Electronic of Germany had advertised. It is based on MGF (I think) 904 and 905 FETs, which are not expensive. They had the board, an aluminum slab for a base, and a parts list. This looked like a good next project for me, not too difficult to step up to 3 gig after 2 gig. It is described in the DUBUS Technik books. It has an extra input stage that would allow it to achieve full output with about 50 mW of drive. Its specifications were just what I needed for the band.

The 5 GHz band starts to get tricky. Construction detail and quality become even more important than on the lower bands. I decided to try a DB6NT design. I had no base plate, so I was going to have to learn simple machining with no experience.

For receive LNAs, I knew that I had to copy an LUA or a KH6CP design in order to get the performance I desired.

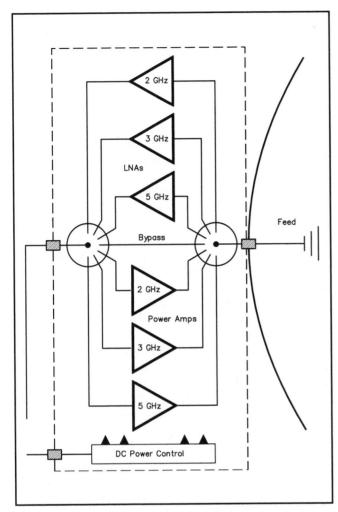

Fig 2—Three-band, back-of-the-dish booster.

Down East Microwave had begun to make kits of the LUA designs (around the ATF-36077). I picked up the 5 gig board and parts, and found a suitable enclosure in the junk box. For the 2 and 3 gig units, I ordered the enclosure from DEM as well.

Acquiring the Parts

Before building anything, it's a good idea to acquire all the parts. In the case of the 2 gig amp, I already had all the parts. For the 3 gig amp, I decided to order the board and the aluminum base plate from Eisch Electronic in Germany. They take credit cards and when I called I found that they would speak English (a lot better than I could speak German!). Eventually, I placed two orders by fax (I didn't include everything on the first order). It took about three weeks to get each order. Because of the added costs, I only ordered the parts that I could not readily obtain in the US. I got most of the standard parts (surface mount electrolytics, integrated circuits, voltage regulator) from Digi-Key, Mouser, and/or Newark. I obtained most critical RF components (surface mount resistors, ATC capacitors and some active devices) from Down East Microwave. Altogether, I spent about $85 on the parts for the 3 gig amp. This cost included overseas shipping and money exchange. Some of the parts used were from my shack supply, so if I had to buy everything new, including the heat sink, it would have been more like $150 for parts.

For the 5 gig amp, I had been keeping my eyes open for about 3 years for devices, and happened to find a supply of Avantek 5964-3, which are 3+ watt devices, and some Mitsubishi MGFC39V-6472A, which are 8+ watt devices. The first could drive the second quite well. Neither of these devices is designed to work at 5760, although the 5964-3 is typically tested as low as 5700. The 6472 is designed to run (as you might assume from its part number) from 6.4 to 7.2 GHz, so I expected that it would need a little tuning to bring it to power at 5760 MHz.

Table 2

Free Advice for Novice Machinists

Item	Free Advice
Placing a hole	Use a fine punch to scribe first, then a larger punch.
Drilling a hole, yes ANY hole	Drill a first pilot with a #55 or the smallest your press will hold, then drill a larger pilot hole, then drill the final hole. Hassle? Yes. Worth it? YES!
. . . continued	Use a tapping or drilling fluid on the bit, such as Aluma-Tap. After each 1/8 inch of depth, clean the bit and add fluid. Your bits will last forever.
Tapping	Use a tapping fluid or wax. Use it every time you insert the tap.
. . . continued	Rotate forward 1/2 to 3/4 turns, and then back 1/4 to 1/2 to break the thread chips. Never force the tap. Always rotate without any sideways force.
. . . continued	Remove the tap every 5 turns or so to clean and re-lubricate.
Cutting	Measure and mark at least twice, cut slightly big and file the edge down.
Goofing up	Take a walk, watch TV, get on the air for a while. Sometimes long projects just need time off in order to clear the mind.

At first I wanted to use a straightforward design by Al Ward for a single stage. It would need to be changed to a two stage and have a bias/regulator power supply added. Since I had not actually made a printed circuit board in about 15 years (and even then I was not working on RF boards), I decided to obtain a DB6NT, two-stage ready-made board as well. This turned out to be a good idea. To make my own board, I tried a laser-printer/iron-on technique from a supplier of a pattern transfer kit. There were many descriptions of what can go wrong and what to do about it. After three tries, I thought I had good transfers (onto an expensive piece of Duroid) and went ahead with etching. The results were not good. Next time I try to make my own RF boards. I will either directly lay tape on the board. or use a photoetch technique. This time, I would use the DB6NT board. That board is two stage, where the first FET is meant to be a much smaller one than the 5964, so I had to squeeze it in.

Parts for the 5 gig amp were obtained from Digi-Key and Down East. There was no baseplate, so I decided to take a simple approach to creating one. This is described below. I obtained metals and hardware (those tiny 080 hex cap screws, tapping tools, etc) from Small Parts, Inc.

The biggest problem that I have with acquiring parts is that I always forget something. Usually this means another minimum order, which can mean $25, even if I only forgot a $0.32 resistor. At first I thought that I would get better at making sure that I would order everything, and perhaps I have become better. Lately I have adopted another technique, and I find this quite satisfactory. I keep a list of items that I would like to have, but don't absolutely need. Then, when I goof and have to place a small order, I first ask my friends if they need any small parts (to pad the order) and if I need more, I order from my "wish list." The other advice I have is to be as generous as possible with those hard to find parts in the stash. Generosity is usually paid back with interest.

Building

Oh my, isn't construction fun! I get both great frustration and satisfaction from building things myself. There are several aspects of building an SHF amplifier that can each be a challenge. The two construction challenges are soldering those tiny things (just where you want them), and the second is drilling, tapping and cutting aluminum (just how you want it).

After years of practice, I am getting to an acceptable level (although far from the best) at soldering—actually making the electronics. The only challenge for the 5760 amp was that there was no description of this amplifier in any of the texts I had from DUBUS. I was either missing the issue, or this particular amp wasn't published. In any case, the only important unknown was the placement of the power supply parts, and it turned out to be a fairly standard design, so that was not a problem.

When it comes to the mechanical parts, however, I am really a novice. Ken, W1RIL, has figured out how to convert a drill press into a small milling machine, appropriate for this kind of work. He has helped me by making a very nice enclosure and base for a 10 GHz amplifier. Someday I

hope I do machining as well as he, but I don't know if I have the patience. I built the 3 GHz amplifier enclosure by bending hobby brass in two "Ls" to surround the aluminum base. The transistors for that amp are shallow enough that slots do not need to be cut in the base. It worked! In the process I learned all about drilling and tapping 6-32, 4-40, 2-56, and yes 0-80 holes. I've broken several taps, drilled some holes in the wrong spot, and made a few cuts at a bad angle or too short. It took me a while to learn some of the following things, that I thought I would summarize to help you if you are also a novice amplifier builder.

I don't doubt that you could learn these and more from a machinist, or in a good high school shop, but somehow I missed that year of shop—we seemed to always be working with wood. I doubt that walnut would make a good amplifier base. Anyway, I'm not the expert, so I offer the following as "free advice" (Table 2) and hopefully it will be worth more than it costs.

Because I have not converted my drill press into a mill, I could not cut slots into the base plate for the transistors in this amp. Instead, I made a second, "false base" of thin aluminum between the board and the real base (see Figure 3). By cutting slots into this false base with a nibbler tool, and drilling it with "through holes" for all the screws that hold down the board, the power FETs could be recessed as needed to keep their leads at the level of the striplines on the board. Well, this was the idea anyway.

Not following my own advice, I calculated the wrong depth for the transistors, and so the 0.032 aluminum sheet was a little too thin to achieve the proper depth. So, the power FET leads had to be bent down to the board approximately 0.02 inch. This sounds small, but I felt pretty lucky that I didn't end up with instability. Two layers of 0.032 would have been just a little too much. I do not recommend this approach unless you absolutely cannot get the slots machined out properly. If you decide to use this approach, try to use material of the right thickness, and "measure twice" with a caliper.

It is often recommended that you use silver loaded epoxy to glue down the board so that a good RF contact is made to the base, resulting in a good match to the transistors and the RF connectors. I agree that this is good advice, but not always needed. At 10 GHz and above, this is probably always the case. At 5 GHz and below. I recommend constructing without the epoxy. Then, when testing, press down hard around the FETs and the connector parts of the board

---

Table 3

Input/Output of the 2-Stage 5760 Amplifier

dBm input	dBm output
16.5	36.5
17.5	37.4
18.5	38.2
19.5	39.0
20.5	39.4

Saturation is reached at 20.5 dBm input, the maximum drive power available.

to see if the contact resistance is contributing to any loss of performance. If it is, then silver epoxy is in order. Otherwise, don't bother. My board had plenty of screws in those critical places, and the epoxy was not needed.

By the way, this epoxy is also quite expensive and has a short shelf life. Unfortunately, it doesn't take a lot of epoxy to glue down a board. Because the material tends to harden in the tube, and any exposure to air will make this worse, I would recommend using the entire epoxy kit at one time. So, to reduce waste, it's a good idea to store up all the projects that need silver epoxy and do them all in one sitting.

Testing

One should always perform a dc test before placing the power FETs into an amp. This is true of LNAs as well. Make sure that the amplifier can handle the power load without significant change in voltage. To do this, temporarily substitute for the FETs power resistors that will draw the same current. Set the bias supplies to –1.5 V. If you have a 'scope, try to make sure that the bias comes on before the $+V_{dd}$ supply. Then install the power FETs. The next step is to set bias. With a load on the input and output of the amp, adjust bias for the right idle current for each device. The big-boy 10 watters can take as much as 2.4 A of drain current at the proper bias setting. This represents over 20 watts of dissipation! Things get hot quickly. Once bias is adjusted, it's time to tune up the amplifier.

On 2 gig amps, it's common to have variable piston capacitors as part of the tuning process. On all stripline solid-state amplifiers (even on amplifiers with tuning capacitors), the best gain and power can be achieved through some "snowflake" tuning. This kind of tuning amounts to placing small pieces of copper foil along the striplines of the power FETs and soldering them in place. It's a process that can take hours, even several sittings, before the output and gain are maximized. Usually, the best results are obtained with two pieces of foil along each section of stripline.

Unfortunately, there is sometimes not enough space to place them exactly right, so three pieces might be used. I have found that in some cases only one piece or rarely, no pieces, are needed. Anyway, here is the process.

Set up a known input level at about 10 dB below the expected input level needed to drive the amp to saturation. It is best to have a dial attenuator in the input line. Attach an output level meter. Use a known (calibrated) directional coupler and a good quality load. If a load with a great return loss at the frequency you are testing isn't available, then a long length of good quality coax with a less-than-excellent load at the end will suffice. Make sure that the coupling factor plus the maximum reading of the power meter exceeds the maximum output of the amplifier—in other words, don't set the system up in a way that might burn out your power head! If there is excess power, then insert attenuator(s) between the coupler and the power head. Now you are set up for gain/power measurement (see Figure 4).

To tune a solid-state microwave amplifier, you need to make a set of snow-flake tools. Find some insulating sticks, such as pieces of Teflon or plastic rod, or tuning tools, or wooden Q-tips, glue different size pieces of copper foil to the flattened tips, ranging from about 0.05 to about 0.15 inch on a side. With the unit powered and RF applied, place different size foil along the striplines to find the "hot spots," where the output jumps up. I usually go from the input end to the output, but end up going back several times to all hot spots anyway. Once you have decided where to place or change a piece of foil, shut off power, solder a best size piece to a hot spot, reapply power and RF, and see what the improvement is. Always keep track of the power level to be sure that things are improving. Try to extend the size of the piece you just added with a small piece. If it improves, shut it off, replace with a bigger piece, and try again. If it does not improve, you can go on to the next hot spot. Sometimes, I find that I can go back after tuning all the hot spots and reduce or enlarge an earlier placed piece to improve the system. During all of this you have to be very, very careful to never short out a bias line. Doing this could easily fry the power FET. Once the tuning is pretty well maxed, increase the RF input power to see how well it performs. It is often

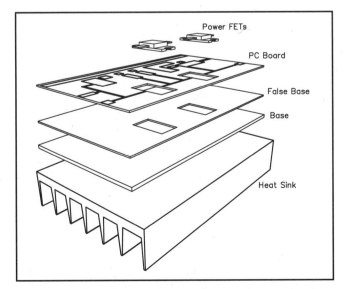

Fig 3—The method used for building the 5 GHz amp.

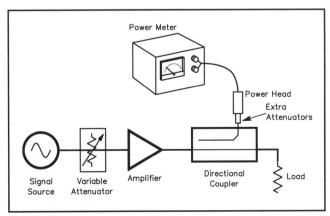

Fig 4—Testing setup.

necessary to readjust the bias at full RF, and then retune the snowflakes for maximum output.

After the amplifier is tuned, I make a chart of the input and output levels, and if there is a power monitor, I record this voltage as well. Although 2 dB increments are sufficient, 1 dB will give a good indication of saturation/power compression. This all helps immediately when determining best input power levels, and later if you want to evaluate an amplifier to determine if there is something failing. I was fortunate, in that the amplifier reached its saturation point at exactly the maximum level that is available at the input to the booster.

I ran each amplifier at full power (key down) for five minutes to determine if there is enough heat sink. Because I wanted these all to fit in a fairly small enclosure, and am also concerned about weight, I used heat sinks that are really too small for the power dissipation needed. I compensated by adding miniature fans to the amps. This kept them all quite cool. Within the enclosure, the air will circulate, and eventually will heat up, so key down time will surely be limited in practical operation. I decided to include a remote temperature sensor in the box to monitor for this condition. Figure 5 shows the components and the box prior to assembly.

In summary, unless you want to use waveguide, the best bet is to mount both power and preamps as close to the feed as possible on all bands 2 GHz and above. I thought that doing this kind of work was strictly for the professionals. Many hams who publish microwave designs are indeed professionals in communications. But many are just hobbyists, like me. If you are willing to learn, you probably can build your own.

References and Sources

"Neatness Counts," talk given by Steven Kostro, N2CEI, with handouts, 1995 Microwave Update Conference.

"Milling with a Drill Press," *Proceedings of the 22nd Eastern VHF/UHF Conference*, 1996, Ken Schofield, W1RIL.

"Microwave FETs," talk given by Al Ward, WB5LUA, with handouts, 1996 Microwave Update Conference.

Eisch Electronic Ulm, Katalog, Inh. Annemarie Eisch-Kafkag, Abt-Ulrich-Str. 16, 89079 Ulm, Germany (Tel 07305-232-8).

Small Parts, Inc., Catalog, 13980 NW 58th Ct, Box 4650, Miami Lakes, FL 33014-0650 (Tel 800-423-9009).

Down East Microwave, 954 Rte 519, Frenchtown, NJ 08825 (Tel 908-996-3584). **http://www.downeastmicrowave.com/**

Digi-Key Corporation, 701 Brooks Ave S, Thief River Falls, MN 56701-2757 (Tel 218-681-6674, toll free order line 800-344-4539), **http://www.digikey.com/**

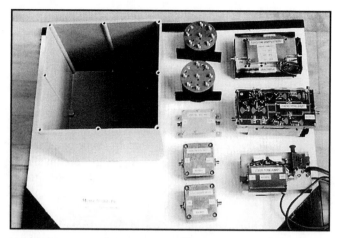

Fig 5—Power amps, LNAs, coaxial relays and box prior to assembly.

A High RF-Performance 10-GHz Band-Pass Filter

By Zack Lau, W1VT
(From *QEX*, July 1997)

Why do I call this a high-performance filter? It offers a very high ratio of Q to size and wastes very little volume in coax transitions or empty space. This is quite important when building a compact portable transverter.

The design is based on one of the very first 10-GHz filters I built—the 2-pole band-pass filter in the TNT, or Tuesday Night Transverter.[1] Unfortunately, the authors omitted two important details—the input/output coupling of the filters and the measured insertion loss. While their design had the potential for great performance, I doubt that many people realized it—even those who built one!

This filter is made from a 1.5-inch section of WR-90 rectangular waveguide. An enclosed cavity is formed by soldering brass sheet stock to the open ends of the waveguide. A pair of posts down the center line creates a pair of coupled cavities. The diameter of the posts determines the coupling—narrower posts increase the coupling between the cavities.

The filter is capable of surprisingly low insertion loss, if you can get the coupling adjusted properly. I chose to use the center conductors of captivated contact SMA connectors as probes. The hard part is adjusting them—just ask anyone who has spent hours filing away! My solution was pretty obvious—use shim stock. After all, isn't this what they sell shim stock for? I made a bunch of spacers of different heights—this allowed me to quickly adjust the height of the coupling probes. With this design, it appears necessary to adjust the probes and resonator tuning for lowest insertion loss. The tuning interacts with the probes, so it didn't appear terribly useful to look at the passband shape when setting the probe coupling. This differs from most other filters, where looking at the passband is an excellent guide.

Using a mixer and isolator to measure the insertion loss, the design shown in Fig 1 has 0.6 dB of insertion loss with a 3-dB bandwidth of 106 MHz. The 144-MHz IF im-age rejection is 33 dB. This should offer excellent performance for receive applications.

For transmit, it might be useful to trade a little insertion loss for better image rejection. With the original design's center coupling using a pair of 3/16-inch posts, the insertion loss increased to 1.3 dB with an image rejection of 47 dB. The 3-dB bandwidth is 36.7 MHz.

Construction

The first step is to get the waveguide, brass metal stock and SMA connectors. Either 2- or 4-hole flange SMA connectors will work, but it is important to use connectors with captivated contacts. Otherwise, the probes will move around, changing the filter alignment. I suggest using brass tubing since it's easier to solder. On the other hand, it might be practical to tap solid brass rod after it's soldered in place—so you can easily mount the filter with screws. Don't forget the clearance required for the 4-40 tuning screws.

As a starting point, I'd try 100-mil probes for the wider filter and 80-mil probes for the narrower filter. These lengths are shortened a bit by the waveguide walls—only 30 to 50 mils actually sticks into the cavity. Adding shim stock reduces the length of the probes, so you might cut the

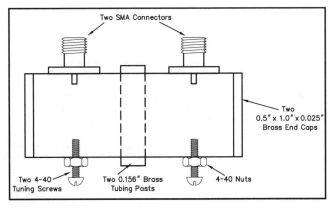

Fig 1—Diagram of the 2-cavity 10-GHz band-pass filter.

[1] Notes appear at the end of this section.

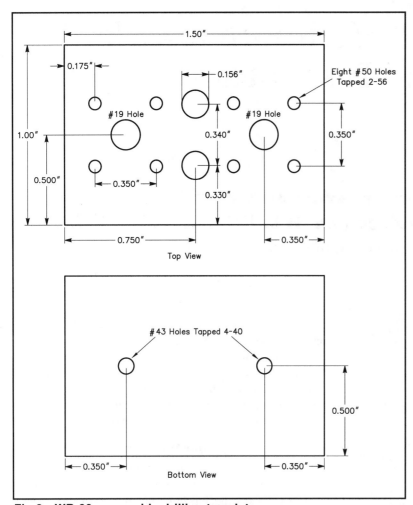

Fig 2—WR-90 waveguide drilling template.

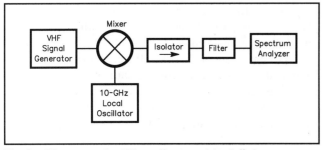

Fig 3—Diagram of a filter alignment test fixture.

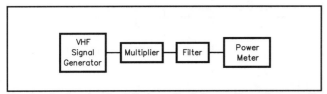

Fig 4—A simpler test fixture that doesn't work as well.

probes long and add shim stock to get the 100-mil starting lengths. The narrower bandwidth version with $^3/_{16}$-inch posts required shorter probes—about 80 mils long.

Once you have the SMA connectors prepared, you can cut and drill the WR-90 waveguide. I filed the cut waveguide edges smooth. If you have 2-hole flange connectors, you need just half of the 2-56 mounting holes shown in the drilling guide. When drilling the holes for the posts, use several drill sizes to get up to the desired hole size. This will help to ensure a tight-fitting round hole that matches the tubing.

After deburring the holes and polishing up the waveguide, I soldered the posts into place. If you do a good job of drilling tight-fitting holes, it shouldn't be difficult to solder the end plates into place without having the posts unsolder themselves, even if you use a propane torch. I use a C-clamp to hold the end plates in place during soldering. I used ordinary 60/40 rosin-core solder.

To tune up the filter, I used a10-GHz mixer/local oscillator to up-convert a signal generator to provide a suitable signal source, as shown in Fig 3. I prefer this technique because I have a spectrum analyzer available to sort out the various mixing products. It is a considerable improvement over the simpler setup shown in Fig 4 using a frequency multiplier and a power meter. The frequency multipliers that I have built are rather frequency sensitive, requiring a calibration plot. Because most filters reflect rather than absorb unwanted signals, the reflections often disturb the operation of mixers and multipliers. The isolator works well with a mixer since all the big signals are near the same frequency. In contrast, a multiplier may have strong signals distributed over a relatively wide frequency range. A resistive attenuator may work better with a multiplier than an isolator designed to work over just an octave.

I estimate that the tuning screws were inserted 0.10 inches into the cavities. Obviously, the exact tuning will vary due to construction tolerances. Adjust the length of the probes and the tuning for minimum insertion loss.

Fig 5 shows a dual-mixer circuit board designed for use at 10 GHz.

It uses 15-mil 5880 Duroid (ε_r=2.2). It's a slight improvement over the one in the June 1993 *QST*, since it uses less board area. This article explains how to tune these mixers for best performance.

A brass frame is soldered around the mixer board using 0.5 × 0.025-inch brass strips. The strips are drilled and tapped to hold SMA connectors. I use 2-hole flange connectors to offset the center pins of the IF connectors so they

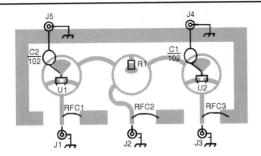

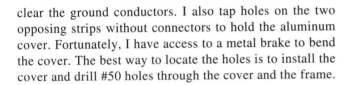

Fig 5—Dual 10-GHz mixer on 5880 Duroid.

C1, C2—0.001 μF, 50 V ceramic disc capacitors.
J1-5, SMA panel-mount female jacks.
R1—51 Ω ¹/₁₀-W chip resistor.
RFC1-3—0.21-inch #28 wires.
U1, U2—HSMS 8202 diode pair.

Fig 6—Full-size etching template for the microstripline mixer.

clear the ground conductors. I also tap holes on the two opposing strips without connectors to hold the aluminum cover. Fortunately, I have access to a metal brake to bend the cover. The best way to locate the holes is to install the cover and drill #50 holes through the cover and the frame.

Then the cover holes are enlarged with a #44 drill to pass the 2-56 screws.

Note
[1]Bailey, Kirk, N7CCB, Larkin, Robert, W7PUA, Oliver, Gary, WA7SHI, "TNT for 10 GHz," *Proceedings of the Microwave Update 1988*.

2 W 10 GHz Amplifier

By Ken Schofield, W1RIL

(From *23rd Eastern VHF/UHF Conference Proceedings*)

The availability of the Microelectronics Technology, Inc Ku band power amplifiers on the surplus market has resulted in a gold mine of parts for 10 GHz operators. This white box not only contains 1 or 2 W (sometimes 1 and 2) IMFET devices but also the V_{dd} and V_g power supplies needed to run them. The 2 W device makes an ideal "afterburner" for the 1 W Qualcom strip available from the West Coast.[2] The box is massive and makes an ideal heat sink for the unit.

Two internal amplifier boxes, complete with covers, contain ceramic substrate gold microstrip. This stuff is so small it requires a jeweler's loupe to even see it. Modifying this gold microstrip is not a task to be undertaken on the amateur workbench! I decided to strip the box, save the screws and other useful parts and make a new board to fit the box—one that I could see, at least partially, without the aid of the jeweler's loupe. See Figure 3.

The new board is mounted in a cut down section of the original amplifier box the cut end of which has been fitted with a new brass plate made from 0.060 stock. A brass carrier cut to the size and shape of the board is fashioned from 0.032 sheet stock. The board and carrier are screwed down to the box bottom using 0-80 ss cap screws previously removed from the box. Before this is done, however, the original device landings in the bottom of the box are removed by milling flush with the rest of the box bottom. You will find that many of the 0-80 screw holes are already in the box bottom. It might be a good idea to leave the hole locations off the board and carrier and custom fit the holes to the box as some boxes may have different hole locations than others. The new hole locations can be added as required.

A schematic of the amplifier is shown in Figure 1. A few comments about the RFCs are in order. These are 0.005-0.006 mil lines. Attempting to get these on the board by etching is an almost impossible task. Strip the insulation from a short length of #20-#22 Teflon-coated stranded wire.

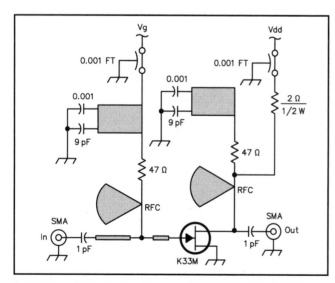

Figure 1—Schematic diagram of 2-W 10-GHz W1RIL amplifier using K33M IMFET.

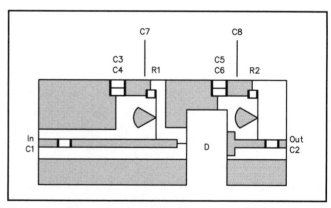

Figure 2—Parts layout of 2-W, 10-GHz W1RIL amplifier.
C1, C2—1-pF 50-mil chip capacitor.
C3, C5—0.001 chip capacitor.
C4, C6—9-pF chip capcacitor.
C7, C8—0.001 FT locations.
D—K33M IMFET.
R1, R2—47-ohm chip resistor.

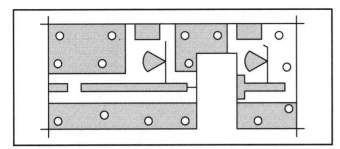

Figure 3—W1RIL circuit board to fit modified Microelectronics Tech Inc amplifier strip.
Board 1.6" × 0.75"
Mtl: Rogers 5880 Duroid, 0.015 mil
er 2.2
50-ohm line = 0.046 wide.

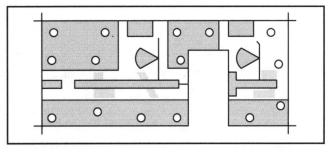

Figure 4—Circuit layout after tuning. Stubs shown in light gray.

Remove one strand of the silver plated wire from the twisted bundle and stretch it slightly to straighten. Solder to the 50 ohm lines and to resistors R1 and R2 keeping the wires straight and flush with the surface of the board. Also solder these lines to the apex end of the quarter circle bypass/decoupling capacitors. These capacitors have a radius of 0.207 and have smooth edges (despite what my computer shows!).

Tuning the amplifier is done by adding stripline pieces—see Figure 4. The gray areas on the 50 ohm line were added to tune amplifier for maximum output into a 50 ohm load. Please note the step in the line at the device input. This is actually a taper (my computer won't do that either!). The line is tapered to the width of the device input gate. Taper the line from 0.046 to 0.022 over $^1/_2$ the distance from the RFC to the device input.

The amplifier gives the following results:

+28 dBm (640 mW) input = +33.4 dBm (2.2 W) output

V_{dd} = 7.48 @ 720 mA. V_g = –1.28

I would like to thank Bruce, N2LIV, for supplying the Tfe board for this project and Don, WB1FKF, for figuring out the power supply capabilities. Without their help I'm sure this project would not have come to fruition.

Notes
[1]Ken Schofield, 21 Forestdale Rd, Paxton, MA 01612, Tel. 508-757-3966.
[2]C. L. Houghton, San Diego Microwave Group.

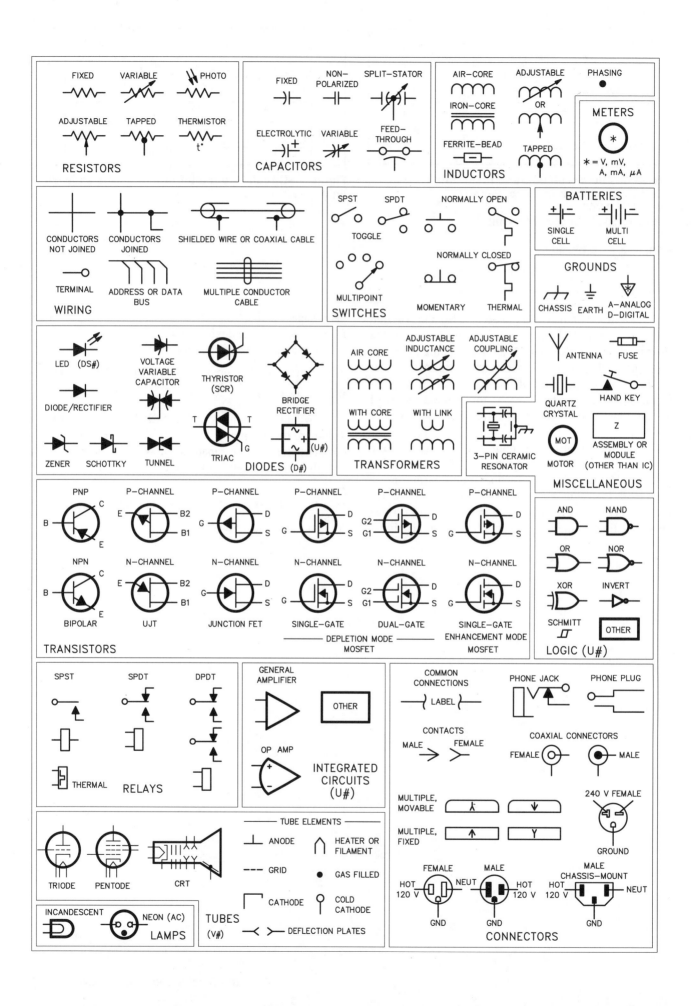

ARRL MEMBERS

This proof of purchase may be used as a $1.50 credit on your next ARRL purchase. Limit one coupon per new membership, renewal or publication ordered from ARRL Headquarters. No other coupon may be used with this coupon. Validate by entering your membership number from your *QST* label below:

UHF/Microwave
Projects Manual
Vol. 2

PROOF OF
PURCHASE

FEEDBACK

Please use this form to give us your comments on this book and what you'd like to see in future editions, or e-mail us at **pubsfdbk@arrl.org** (publications feedback). If you use e-mail, please include your name, call, e-mail address and the book title, edition and printing in the body of your message. Also indicate whether or not you are an ARRL member.

Where did you purchase this book?
☐ From ARRL directly   ☐ From an ARRL dealer

Is there a dealer who carries ARRL publications within:
☐ 5 miles   ☐ 15 miles   ☐ 30 miles of your location?   ☐ Not sure.

License class:
☐ Novice   ☐ Technician   ☐ Technician Plus   ☐ General   ☐ Advanced   ☐ Amateur Extra

Name _____   ARRL member? ☐ Yes ☐ No

_____ Call Sign _____

Daytime Phone (   ) _____ Age _____

Address _____

City, State/Province, ZIP/Postal Code _____

If licensed, how long? _____

Other hobbies_____

_____

Occupation _____

For ARRL use only	MWPROJ2
Edition	1 2 3 4 5 6 7 8 9 10 11 12
Printing	1 2 3 4 5 6 7 8 9 10 11 12

From _____

_____

_____

Please affix postage. Post Office will not deliver without postage.

EDITOR, MICROWAVE PROJECTS MANUAL VOL 2
AMERICAN RADIO RELAY LEAGUE
225 MAIN STREET
NEWINGTON CT 06111-1494

— — — — — — — — — — — — — — — — please fold and tape — — — — — — — — — — — — — — — — — —